Lecture Notes in Mathematics

Edited by A. Dold and B. Eckmann

515

R.M. Miura (Ed.)

Bäcklund Transformations, the Inverse Scattering Method, Solitons, and Their Applications

Proceedings of the NSF Research Workshop on Contact Transformations, held in Nashville, Tennessee, 1974

Springer-Verlag Berlin Heidelberg GmbH

Editor

Robert M. Miura
Department of Applied Mathematics
University of Washington
Seattle, WA 98195, USA

1st Edition 1976
2nd Printing 1989

AMS Subject Classifications (1970): 34-02, 34B25, 34J10, 35-02, 35A25, 35B10, 35C05, 35F25, 35G25, 42A76, 49G99, 58A15, 70H15, 76B25, 78A40, 81A45

ISBN 978-3-540-07687-2 ISBN 978-3-540-38220-1 (eBook)
DOI 10.1007/978-3-540-38220-1

Originally published by Springer-Verlag Berlin Heidelberg New York in 1976

2141/3140-543210

PREFACE

An "NSF Research Workshop on Contact Transformations" was held at Vanderbilt University in Nashville, Tennessee on September 27-29, 1974. The main emphasis of the Workshop was on Bäcklund transformations, the inverse-scattering method, and solitons and how these topics could be applied to the study of various nonlinear partial differential equations of physical interest. These research areas have developed rapidly over the past five years and one of the purposes of this Workshop was to bring together some of the most active researchers to disseminate their results and ideas as well as to find areas of common interest and overlap. The participants (see the participants list on page V) included engineers, physicists, and mathematicians with interests in nonlinear partial differential equations. There were 22 researchers from the United States, two from Canada, and one from Japan.

The Workshop program contained both expository and technical talks and there were numerous informal discussions. This collection of papers represents expanded versions of most of these talks and include many additional details and results not presented at the Workshop. (The paper by Alan C. Newell, who was unable to attend due to the imminent arrival of a new member to his family, was presented for him by the Editor.)

I am particularly pleased to thank the authors of these papers for the their hard work and cooperation in preparing the manuscripts and for their generous patience in waiting for this collection to appear. I also wish to thank the National Science Foundation for financial support of this Workshop under NSF Grant MPS 74-21147. Thanks are also due to Carlene Mathis for her excellent typing of the photo-ready copy of the manuscript. Finally, I wish to extend my appreciation to Gerald B. Whitham of Cal Tech and Walter Kaufmann-Bühler of Springer-Verlag for their interest in getting these Proceedings in print.

Robert M. Miura
Vancouver, B.C., Canada
December 1975

TABLE OF CONTENTS

RESEARCH WORKSHOP PARTICIPANTS

BERRYMAN, JAMES G.
Mathematics Research Center
University of Wisconsin
Madison, Wisconsin 53706

CHEN, HSING-HEN
Department of Physics and Astronomy
University of Maryland
College Park, Maryland 20742

CHU, FLORA YING FUN
Department of Electrical Engineering
Massachusetts Institute of Technology
Cambridge, Massachusetts 02139

CONLEY, CHARLES C.
Department of Mathematics
University of Wisconsin
Madison, Wisconsin 53706

COPE, DAVIS
Department of Mathematics
Vanderbilt University
Nashville, Tennessee 37235

CORONES, JAMES
Department of Mathematics
Iowa State University
Ames, Iowa 50010

ESTABROOK, FRANK B.
Jet Propulsion Laboratory
California Institute of Technology
4800 Oak Grove Drive
Pasadena, California 91103

FARRINGTON, TED
Department of Mathematics
Clarkson College of Technology
Potsdam, New York 13676

FLASCHKA, HERMANN
Department of Mathematics
University of Arizona
Tucson, Arizona 85721

GERBER, PORTER DEAN
IBM Corporation
Thomas J. Watson Research Center
P.O. Box 218
Yorktown Heights, New York 10598

GREENE, JOHN M.
Princeton Plasma Physics Laboratory
P.O. Box 451
Princeton, New Jersey 08540

HIROTA, RYOGO
Department of Mathematics and Physics
Ritsumeikan University
Kitamachi 28-1, Tooji-in
Kita-ku, Kyoto
603 Japan

KAUP, DAVID J.
Department of Mathematics
Clarkson College of Technology
Potsdam, New York 13676

LAMB, GEORGE L. JR.
Department of Mathematics
University of Arizona
Tucson, Arizona 85721

LONNGREN, KARL E.
Department of Electrical Engineering
University of Iowa
Iowa City, Iowa 52242

MCLAUGHLIN, DAVID W.
Department of Mathematics
University of Arizona
Tucson, Arizona 85721

MIURA, ROBERT M.
Department of Mathematics
Vanderbilt University
Nashville, Tennessee 37235

RANGER, KEITH
Department of Mathematics
University of Toronto
Toronto 5, Ontario, Canada

ROGERS, COLIN
Department of Mathematics
University of Western Ontario
London, Ontario, Canada

RUND, HANNO
Department of Mathematics
University of Arizona
Tucson, Arizona 85721

SCOTT, ALWYN C.
Department of Electrical Engineering
University of Wisconsin
Madison, Wisconsin 53706

SEGUR, HARVEY
Department of Mathematics
Clarkson College of Technology
Potsdam, New York 13676

TAPPERT, FREDERICK
Courant Institute of Mathematical Sciences
251 Mercer Street
New York, New York 10012

VARLEY, ERIC
Center for the Application of Mathematics
4 W. 4th Street
Lehigh University
Bethlehem, Pennsylvania 18015

WAHLQUIST, HUGO D.
Jet Propulsion Laboratory
California Institute of Technology
4800 Oak Grove Drive
Pasadena, California 91103

ZABUSKY, NORMAN J.
Department of Mathematics
University of Pittsburgh
Pittsburgh, Pennsylvania 15260

INTRODUCTION*

Robert M. Miura†

Department of Mathematics
Vanderbilt University
Nashville, Tennessee 37235

The study of nonlinear partial differential equations has had a sporadic history up through the present time. In spite of the fact that physical phenomena are crying out for the solution of the underlying nonlinear model equations, few general methods of solution have been devised. Nonlinear partial differential equations exhibiting wave phenomena can essentially be classified as hyperbolic or dispersive (see Whitham [5]). Whereas the theory of hyperbolic partial differential equations is fairly well developed, the theory of nonlinear dispersive wave equations is not well developed. Prototypes of dispersive equations are the Korteweg-deVries (KdV) equation, the modified Korteweg-deVries (MKdV) equation, the nonlinear Schrödinger equation, and the sine-Gordon equation.

The applications which traditionally received the most attention were in fluid dynamics. Recently, however, applications of model equations to nonlinear phenomena in other disciplines are receiving more attention and there is a definite need for more general solution techniques. Some of the applications are to water waves, crystal optics, quantum mechanics, lattice dynamics, active transmission lines, various areas of continuum mechanics, and nerve pulse propagation.

Theoretical progress on these model equations has depended mainly on how rapidly one can generate numerical and approximate solutions which sample as much of the corresponding parameter spaces as possible. For the most part, numerical solutions are a poor means of sampling parameter spaces to extract the qualitative behavior of solutions and, in general, the accuracy of approximate solutions depends on the parameters having small or large values. Furthermore,

*Supported in part by the National Science Foundation under NSF Grant GP-34319.

†On leave at the Department of Mathematics, University of British Columbia, Vancouver, B.C., Canada, V6T 1W5.

aside from linearizations, results have been obtained primarily from "nonlinear perturbation theory." Some of the techniques described in the papers in this collection do not have these limitations but are limited in other ways, e.g. to the types of equations to which they can be applied.

In initial studies of model equations, one looks for special solutions and nonlinear dispersive wave equations are no exceptions. However, here the solutions which consist of steady progressing waves play a special role in the general solutions to the initial-value problems. The solitary wave solutions for these particular equations manifest themselves as "solitons." A solitary wave solution is characterized by being a localized wave pulse which does not change its shape as it moves at constant speed. We include in such a classification, functions which go from one constant value as $x \to -\infty$ to another constant value as $x \to \infty$, but with derivative which is a localized wave pulse. Now at some initial time, consider the superposition of two such solutions with the pulses well separated and each with a different wave speed. The pulses are placed relative to each other such that as $t \to \infty$ they will run into each other. They are called <u>solitons</u> if after the nonlinear interaction they emerge unchanged in wave shape, but can possibly be shifted in position from where they would have been had no interaction occurred. For general initial conditions, as $t \to \infty$, the solitons emerge as distinct entities and form an integral part of the solutions.

In the last 10 years, a number of these nonlinear partial differential equations have been solved by application of an "inverse scattering method." To describe this method in outline, consider a given nonlinear partial differential equation in one space dimension with specified initial data. (A detailed development of this method as applied to the KdV equation is presented in [2].) The inverse scattering method consists of first finding an appropriate associated linear scattering problem (in one space dimension) in which the unknown solution of the given differential equation appears as a potential and the time occurs as a parameter. Then the objective is to construct the potential from the "scattering data." To bring in the time evolution, one uses the specified initial data to determine the scattering data at the initial time and then linear evolution equations for the scattering data are used to determine the scattering data at later times from which the "potential" (solution of the given problem) is determined.

It is now clear that in the study of nonlinear dispersive wave equations, two important research problems are to find soliton solutions and an inverse scattering method. The Bäcklund transformation (BT) is a possible solution to each of these problems. However, it remains to determine if the problem of finding a BT is not as difficult as these original problems.

There is no generally accepted definition of a BT. To describe it in some limited cases, consider a second-order partial differential equation. The BT consists of a pair of first-order partial differential equations relating a solution of the given second-order equation to another solution of the same equation or to a solution of another equation. In the pair of first-order equations, one involves only x-derivative terms and the other involves only t-derivative terms. Although, in general, solution of these first-order equations is also difficult, the Theorem of Permutability provides a method for obtaining new solutions from known solutions without the use of quadratures.

As already mentioned, these areas of research have important applications and this collection contains basic expository and research papers which form an introduction to these subjects and carry the reader to the frontiers of research. The references cited form an important part of the papers and collectively represent most of what has been written on these subjects. Whitham [5] gives an excellent introduction to the field of nonlinear wave propagation. Other expository and research papers are collected together in Leibovich and Seebass [1], Moser [3], and Newell [4]. Forthcoming is a collection of papers on the theory and applications of solitons [6].

The papers collected here deal mainly with three topics: i) Bäcklund transformations, ii) the inverse scattering method, and iii) solitons. The papers range in content from experiments on nonlinear dispersive transmission lines to the use of exterior differential forms. Ironically, this collection appears exactly 100 years after Bäcklund's first paper on his transformation theory which appeared in 1875. We now briefly outline the contents of the papers to help guide the reader. With the exception of the first three papers by Lonngren, Chu, and Hirota and the last paper coauthored by Flaschka and McLaughlin, the papers are collected in the order in which they were presented at the Workshop.

The first two papers by Lonngren and Chu treat experimental situations, a nonlinear dispersive transmission line and stimulated Raman and Brillouin scattering, respectively, in which soliton phenomena are observed. Beginning with a discrete nonlinear transmission line, Lonngren derives the model equations and finds solitary wave solutions obtainable in the experiments. Soliton behavior has been observed but the analytical work remains incomplete.

On the other hand, Chu derives both the model equations and the equations for the inverse scattering method and finds the soliton solutions. However, direct comparison with the experimental situation is not possible because of the coordinates chosen and the unrealistic initial conditions used. It is an open problem to modify these results to correctly take these into account.

Hirota has many contributions to this area and he presents here a direct method for finding exact solutions of a number of different nonlinear evolution equations. His procedure is to replace the dependent variable(s) by a ratio of functions which satisfy coupled bilinear differential equations. (This is reminiscent of the use of Padé approximants.) The form of the equations is simplified by introducing new operators, $\partial/\partial t \rightarrow \partial/\partial t - \partial/\partial t'$, $\partial/\partial x \rightarrow \partial/\partial x - \partial/\partial x'$, in an extended space of four variables, letting the dependent variables depend on these extended variables in the differential equation, and then restricting $x = x'$, $t = t'$. The method then is to expand the numerator and denominator in the ratio as series in a parameter ϵ and to evaluate the coefficients by the usual perturbation series method. For the case of solitons, these series reduce to finite sums and give explicit formulas for the solutions. Some of the equations for which soliton solutions are obtained include the modified Korteweg-deVries equation, the nonlinear Schrödinger equation, the two-dimensional Korteweg-deVries equation, and the two-dimensional sine-Gordon equation. One of the outstanding problems in this area is the extension of results to higher dimensions. Hirota's method is a step in this direction.

The remaining papers deal principally with Bäcklund transformations and their connection with the inverse scattering method and solitons.

Lamb gives some of the historical background dating back to the original research by Bäcklund on pseudospherical surfaces. He illustrates the form of the BT and the use of the Theorem of Permutability for finding solutions of the sine-Gordon equation which was originally derived for the problem of pseudospherical surfaces. One direct way of deriving BT is due to Clairin. Lamb outlines the procedure and then gives a detailed derivation of the BT relating solutions of Liouville's equation and the wave equation. Since the general solution of the wave equation is known, this leads to the general solution of Liouville's equation. An extensive list of references to earlier works is included.

The next two papers by Scott and Rogers give extensive applications of BT to various physical problems. Some of the areas discussed are Josephson junctions and transmission lines, wave propagation in active nerve fibers, nonlinear optics, ion-sound waves, gasdynamics, magnetogasdynamics, elasticity, viscoelasticity, and nonlinear filtration.

Scott finds the BT for the linear Klein-Gordon equation in polar coordinates which is used to successively generate the radial eigenfunctions. For nonlinear diffusion equations, he shows how the BT can generate traveling wave solutions. For Burgers equation, it becomes clear that knowing the BT may not be as useful as knowing the linearizing transformation, in this case the Hopf-Cole transformation of Burgers equation to the linear diffusion equation. Scott gives extensive discussions of various nonlinear Klein-Gordon equations in one and two space dimensions which arise in problems associated with Josephson junctions and Josephson junction transmission lines. The BT generates the soliton solutions and in these applications the soliton represents a quantum of magnetic flux and the N-soliton solutions represent propagating bundles of magnetic flux. For the two-dimensional sine-Gordon equation, the BT has not been found, but an attempt in this direction is to consider the nonlinear Klein-Gordon equation with a sawtooth shaped nonlinear term. Finally, a Boussinesq equation is derived for ion sound waves in a plasma. Hirota has found an N-soliton solution and Hirota and Chen have found the BT.

Rogers gives a comprehensive introduction to applications of generalized BT to a variety of areas of continuum mechanics and gives an extensive list of references to the literature. Generalized BT allow a vector-valued dependent variable in place of a scalar valued one. The discussion is confined to the case of two independent variables but this still leaves a wide range of applications. A definite limitation is the application to systems of linear first-order partial differential equations. Work on nonlinear systems using these generalized BT remains to be done. The general theory which includes dependence on the independent variables is developed in matrix notation. A particular case of physical interest is the reduction of the hodograph equations of gasdynamics to canonical form for subsonic, transonic, and supersonic flow. The Stokes-Beltrami system can be solved by repeated iteration of the matrix transformations. Application of these ideas to problems in magnetogasdynamics and in elasticity illustrate using the hodograph transformation followed by a matrix Bäcklund-type transformation to yield either the Cauchy-Riemann equations or the wave equation.

The three papers by Estabrook, Wahlquist, and Corones and Testa expound on the uses of ideas from the calculus of exterior differential forms, of pseudopotentials and prolongation structures for studying nonlinear partial differential equations, and of connections with other areas in the theory, notably, BT, conservation laws, and the inverse scattering method.

Estabrook provides an introduction to the algebra and calculus of differential forms and lays the foundation for the differential form methods which he and Wahlquist have used for studying partial differential equations. This paper gives an introduction to both the ideas and the vocabulary used. Some of the concepts discussed are n-dimensional differentiable manifolds, p-forms, vectors, Lie differentiation, solutions of partial differential equations as integral manifolds of a set of differential forms, similarity solutions, conservation laws, pseudopotentials, and prolongation structures.

Wahlquist illustrates the use of potentials and pseudopotentials on the KdV equation. The connection with the inverse scattering method is shown. Comparison of the BT for the KdV equation, found originally by Wahlquist and Estabrook, with the equations governing the pseudopotentials obtained from the prolongation structure shows that the pseudopotentials can be interpreted as the difference of two solutions related by the BT. It is shown how to generate an infinite hierarchy of solutions by finding the corresponding transformations of the pseudopotentials using permutation symmetry on the BT. This generalizes the hierarchy of multisoliton solutions since one can begin with any solution of the KdV equation. Beginning with the general steady progressing wave solutions, of which the cnoidal wave and solitary wave are special cases, the corresponding pseudopotential is obtained. Beginning with the cnoidal wave, the transformed solution appears as a superposition of three basic types of waves: i) cnoidal waves, ii) a modulated soliton, and iii) spatially damped oscillations. The cases starting with a cnoidal wave and a solitary wave are investigated.

Corones and Testa present the introductory stage of their investigation of the uses of the pseudopotential for constructing BT, conservation laws and finding the associated inverse scattering problem. They review the differential forms approach of Wahlquist and Estabrook leading to the construction of pseudopotentials and the prolongation structure associated with the original partial differential equation. They consider the case of one pseudopotential and present an apparently different method for obtaining BT, at least it is a labor saving method. One can determine if there is any possible associated first-order linear eigenvalue problem needed for the inverse scattering method if there exists a prolongation structure with functions which are linear in the pseudopotentials. Such linear prolongation structures are found for the Hirota equation and the Burgers-modified KdV equation. However, the existence of such a linear prolongation structure does not guarantee the existence of an eigenvalue problem. This remains an open problem.

The paper by Rund represents a deviation from the approaches taken thus far. He considers a pair of partial differential equations $E(x) = 0$ and $D(y) = 0$ in the unknowns x and y. Then a system of one or more relations in x, y and their derivatives is called a BT if they insure that $D(y) = 0$ if $E(x) = 0$ and conversely. The partial differential equations considered are assumed to be Euler-Lagrange equations derived from a variational principle with Lagrangian L. He defines a variational BT between x and y as a relationship such that the difference $L(y) - L(x)$ is a derivative. The variational BT is contrasted with the simple BT which make the difference $E(y) - E(x) = 0$. It is shown that a simple BT need not be a variational BT. A strong BT is a simple BT which implies both $E(x) = 0$ and $E(y) = 0$. The variational theory leading to the definition of the variational BT is developed and applied to a general class of sine-Gordon type equations. For the two-dimensional case, the class of equations admits BT only if the nonlinear terms f satisfy the restriction $f'' = kf$, a condition found earlier by Kruskal for the existence of infinitely many conservation laws and by McLaughlin and Scott for the existence of a BT using a different definition. The simple BT for this class of equations are also variational BT. Variational BT are found for the KdV equation and the MKdV equation. For the quartic nonlinear MKdV equation, however, there is a simple BT but not a variational one. Rund also shows that the idea of a simple BT is useful even when there is no underlying variational principle. He derives a strong BT for Burgers equation yielding the Hopf-Cole transformation. Finally, he shows that there is a simple BT relating the KdV equation and generalizations of the KdV equation, but it is a strong BT only when relating the KdV equation and the MKdV equation.

The last three papers by Newell, Chen, and Flaschka and McLaughlin deal with relationships between BT and the inverse scattering method. This is one of the most exciting possible uses of BT since at present we have no systematic way of starting with a given nonlinear partial differential equation and then generating the associated linear equations for applying the inverse scattering method. It should be pointed out that up to the present time, various ad hoc procedures have been used to find the associated linear equations and the path from the BT to the inverse scattering problem has not been found until after the inverse problem was already known. Newell and Chen both concentrate on going from the inverse problem to the BT. In addition, the paper by Flaschka and McLaughlin shows relations with the ideas of canonical transformations from Hamiltonian mechanics.

Newell begins with a generalization of the Zakharov and Shabat eigenvalue problem and the associated linear time-evolution equations and states the class of evolution equations (developed by the Clarkson group) which can be treated by this eigenvalue problem. From the linear equations, a change of variables leads to a pair of coupled Riccati equations which in turn lead to the BT. The reverse route from the Riccati equations to the linear equations is possible under additional restrictions and some of these ideas may lead to a general procedure for finding the associated linear problems. By restricting the variables arising in the class of evolution equations, one fixes the x-component of the BT. However, there still remains a great deal of freedom in the choice of the t-component of the BT which is determined once the specific evolution equation is chosen. A number of different examples are given including the sine-Gordon equation, the MKdV equation, the sinh-Gordon equation, and the KdV equation. Newell shows how the transformation relating solutions of the KdV equation and the MKdV equation in fact relates two infinite families of equations, one family of which has the time-independent Schrödinger equation as the associated linear eigenvalue problem. Special solutions of these two families of equations are studied and, in particular, the solitary wave solutions and the similarity solutions are examined.

Chen gives a detailed derivation of the BT for the KdV equation starting from the associated linear eigenvalue problem and linear evolution equation. For the class of evolution equations developed by the Clarkson group, he shows how one can use a simple gauge-like invariance transformation to derive the BT. Three separate classes of equations within the Clarkson group category are used as examples and for two of these classes, two equivalent forms of the BT are derived. The BT corresponding to higher-order scattering problems can be obtained and Chen illustrates this for the Boussinesq equation.

The final paper by Flaschka and McLaughlin is concerned mainly with BT and the Toda lattice problem. This is the only paper in the collection which concerns itself with the discrete problem. For the BT the emphasis is not on the transformation from one solution of an equation to another solution of the same equation, but rather on the transformation of one scattering problem to another scattering problem. Thus the point of view is directed towards the spectral theory of Sturm-Liouville problems. The ideas are developed for the KdV equation with its associated eigenvalue problem. The BT is viewed as a transformation of coefficients for this eigenvalue problem. From this point of view, the evolution equation is unimportant and one is dealing principally with the x-component of the BT. The basic formula used is that relating the eigenfunctions of the two related eigenvalue problems. In particular, this formula determines how the scattering data are transformed. Since the various pieces of the scattering data have direct interpretations with respect to the solutions, such as solitons and their location, one can ascertain what effect the BT has on certain solutions without direct computation of the solution. It is found that the BT will either add a soliton and/or shift the phase of the continuous spectrum. When the t-component of the BT is taken into account, it is found that the BT commutes with the KdV flow.

When the original solution of the KdV equation is a cnoidal wave and the new solution is to be periodic, the needed BT does not add solitons. In terms of the spectral characterization of these periodic solutions, the BT does not open up any new gaps in the spectrum of a periodic Sturm-Liouville operator. To add a soliton means to add an eigenvalue to the continuous spectrum. This leads to a nonlocal perturbation of the original periodic potential. The connection between the KdV equation and the linear eigenvalue problem has led to reformulating the problems in the framework of Hamiltonian mechanics. The ideas of Poisson brackets, canonical transformations, and constants in involution have been used to advantage in further development of the theory. For example, the BT is a canonical transformation on the set of periodic potentials with constant mean value.

The Toda lattice problem is recast in canonical variables and then a discrete version of the inverse scattering method is carried out to solve the initial-value problem. The remainder of the paper discusses various aspects of the canonical setting and its consequences. The two classes of motion invariants, the "action variables" and the "usual" constants, are shown to be equivalent and their interpretations are given. An adaptation of the method used by the Clarkson group for generating a class of equations solvable by the inverse scattering method is used to derive a class of completely integrable systems. Comparison of the Toda lattice with the harmonic lattice leads to an analogous definition of normal modes for the Toda lattice.

REFERENCES

[1] S. LEIBOVICH AND A.R. SEEBASS, Eds., Nonlinear Waves, Cornell University Press, Ithaca, N.Y., 1974.

[2] R.M. MIURA, The Korteweg-deVries equation: A survey of results, SIAM Rev., to be published.

[3] J. MOSER, Ed., Dynamical Systems, Theory and Applications, Battelle Seattle 1974 Rencontres, Springer-Verlag, New York, N.Y., 1975.

[4] A.C. NEWELL, Ed., Nonlinear Wave Motion, American Mathematical Society, Providence, R.I., 1974.

[5] G.B. WHITHAM, Linear and Nonlinear Waves, John Wiley and Sons, New York, N.Y., 1974.

[6] Proceedings of the Conference on the Theory and Application of Solitons, January 5-10, 1976, Tucson, Arizona, Rocky Mountain J. Math., to be published.

EXPERIMENTS ON SOLITARY WAVES*

Karl E. Lonngren

Department of Electrical Engineering
The University of Iowa
Iowa City, Iowa 52242

I. INTRODUCTION

This paper reviews several experiments performed on a nonlinear dispersive transmission line which was constructed at The University of Iowa in order to illustrate properties of solitary waves and "solitons." In addition, the shape of the solitary wave which propagates along the line is predicted theoretically. A number of the results described here have appeared in print elsewhere [1] - [5].

In Section II, the nonlinear dispersive transmission line is described and its nonlinear and dispersive properties are discussed. To our knowledge, only two other experiments on solitary waves on transmission lines have been reported. Hirota and Suzuki [6], [7] constructed a 900 section line with each section consisting of a series inductor and a shunt nonlinear capacitor. Gorshkov et. al. [8], [9] used a circuit similar to the one we will describe, but their's was slightly more complicated. The equation governing the traveling wave solution is also derived.

In Section III, we obtain the solitary wave solutions. The equation which we have derived is an additional one to those listed in the excellent review paper on solitons by Scott, Chu, and McLaughlin [10].

In Section IV, we illustrate several properties of solitary waves which were ascertained from experiments performed on this transmission line. Section V is the conclusion.

---

* Supported in part by the National Science Foundation, Grant No. ENG74-00704.

II. EXPERIMENTAL TRANSMISSION LINE

A typical section of the 50-section line is shown in Figure 1. Each section consists of a parallel resonant circuit $(\omega_o/2\pi = 1/2\pi\sqrt{LC_S} \sim 30$ MHz) in the series branch and a reverse biased p-n junction diode (Western Electric F54837 diode) whose capacitance is a nonlinear function of the bias and the signal voltages in the shunt branch. We found in our experiment that the diode

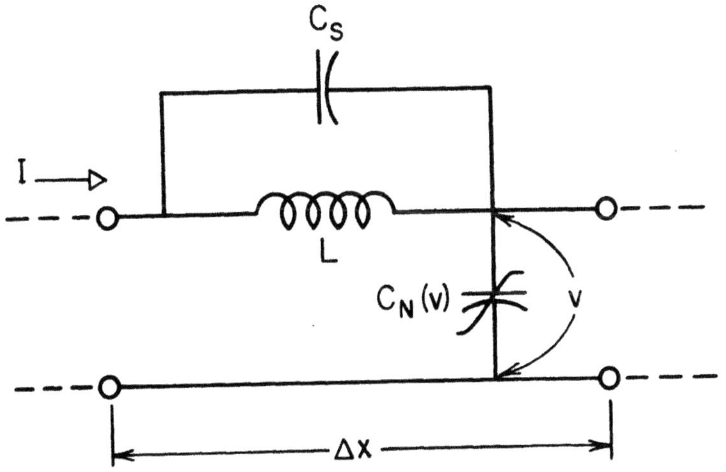

Figure 1. A typical section of the nonlinear dispersive transmission line. In the experiment: $L = 0.14$ micro henries; $C_S = 221$ pico farads; $C_N(V) = (130$ to $500)$ pico farads (Western Electric #F 54837); and $\Delta x = 2$ cm.

could be described by the expression $C_N(V) = C_{NO}(V/\bar{V})^{-n}$ where $\bar{V}$ is a normalizing constant and $n \sim 1/3$. It is also possible to use this line to simulate several plasma wave experiments where solitons and shocks have been observed [11].

From Figure 1, we can write

$$I(x) - I(x - \Delta x) = \frac{\partial}{\partial t} [VC_N(V)\ \Delta x] ,$$

(1)
$$V(x) - V(x - \Delta x) = L\Delta x \frac{\partial \hat{I}}{\partial t} ,$$

$$V(x) - V(x - \Delta x) = \frac{\Delta x}{C_S} \int (I - \hat{I})\ dt ,$$

where we have assumed that the current $\hat{I}$ passes through the inductor L and the current $(I - \hat{I})$ passes through the linear capacitor C_S. The three expressions describe the change of the current I along the line due to some current being shunted through the nonlinear capacitor $C_N(V)$ and the change of voltage V along the line caused by the current $\hat{I}$ in the inductor L and $(I - \hat{I})$ in the linear capacitor C_S respectively. In the "long wavelength" approximation (or $\lim \Delta x \to 0$), the set of equations in (1) becomes

$$\frac{\partial I}{\partial x} = \frac{\partial}{\partial t} [VC_N(V)] ,$$

(2)
$$\frac{\partial V}{\partial x} = L \frac{\partial \hat{I}}{\partial t} ,$$

$$\frac{\partial^2 V}{\partial x \partial t} = \frac{1}{C_S} (I - \hat{I}) ,$$

where V, I, and $\hat{I}$ are functions of x and t and we have eliminated the time integral by differentiation. The initial condition for these equations are specified by the properties of the elements, i.e., the current through an inductor and the voltages across the capacitors cannot change instantaneously.

At this stage, we shall be more interested in examining the traveling wave solution rather than the initial-value problem. In (2), we assume that $I = I(x - at)$, $\hat{I} = \hat{I}(x - at)$, and $V = V(x - at)$ and denote differentiation with respect to $(x - at)$ by a prime. Equation (2) becomes

(3a)
$$I' = - a[VC_N(V)]' ,$$

(3b)
$$V' = - aL\hat{I}' ,$$

(3c) $$- aV'' = \frac{1}{C_S} (I - \hat{I}) .$$

From (3a) and (3b), we integrate and find

(4) $$I = - a[VC_N(V)] + \alpha ,$$
$$V = - aL\hat{I} + \beta ,$$

where α and β are constants of integration. We now substitute (4) into (3c)

(5) $$V'' + \frac{1}{aC_S}\left[\frac{V}{aL} - aVC_N(V)\right] = - \frac{1}{aC_S}\left[\alpha - \frac{\beta}{aL}\right].$$

The dispersion relation for small signal propagation can be found from the homogeneous part of (5) under the assumption that $C_N(V) \sim C_{NO}$ and $V(x - at) = V_0\, e^{i(kx-\omega t)}$ where $a \equiv \frac{\omega}{k}$. From (5), we write

(6) $$\frac{\omega}{k} = \pm \frac{1}{\sqrt{LC_S k^2 + LC_{NO}}} .$$

In the long wavelength case, $k \equiv 2\pi/\lambda \to 0$

(7) $$\frac{\omega}{k} \sim \pm \frac{1}{\sqrt{LC_{NO}}} .$$

Experimentally measured dispersion curves obtained for different values of bias voltage V_{bias} are shown in Figure 2. These curves were obtained by measuring the wavelength of the standing wave set up by an A-C short at one end (blocking capacitor needed to apply V_{bias}) when a very small ($\sim$0.1 volt) variable frequency sine-wave signal was applied at the other end.

In addition to being dispersive, the line is nonlinear. This is demonstrated by examining the response of the line to positive and negative pulses of different amplitude (pulse duration < $1/\omega_0$ where ω_0 is the resonant

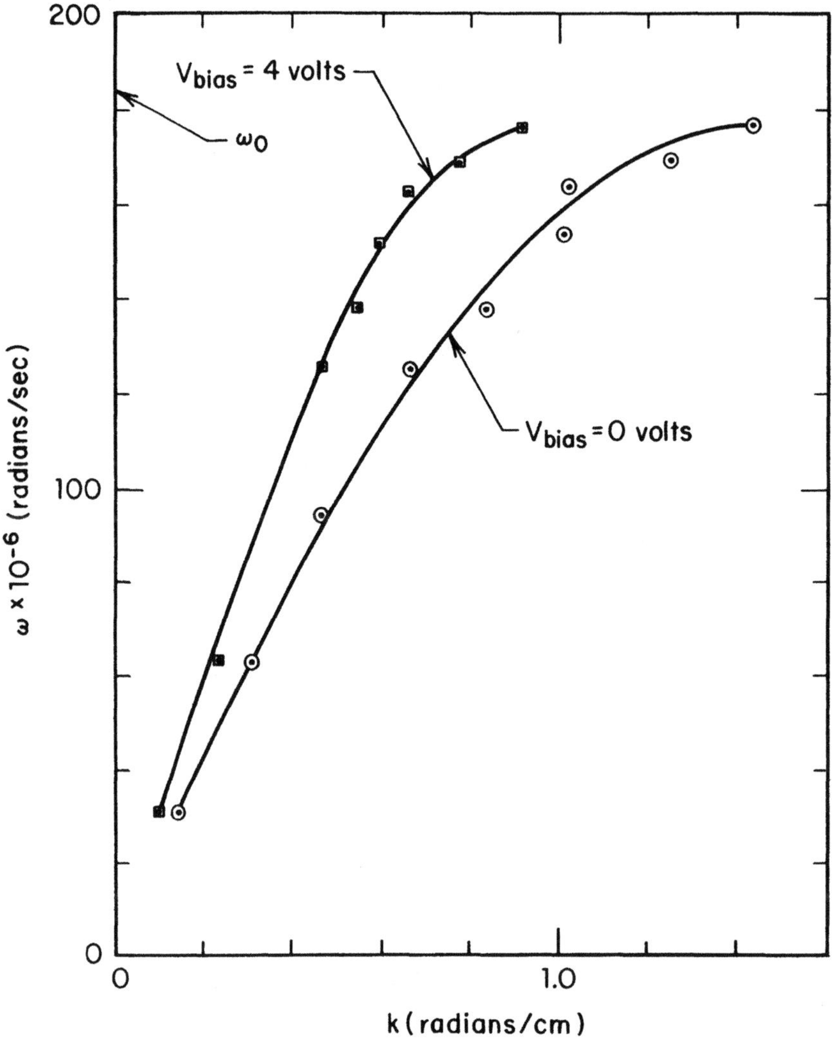

Figure 2. Experimentally measured dispersion curves for the transmission line as a function of the bias voltage applied to the nonlinear capacitor.

frequency of the linear parallel resonant circuit of the series branch) with fixed reverse bias (V_{bias} > 3 V). Typical results are shown in Figure 3 where we have observed the response of a 30-ns pulse at a point 50 cm from the point of excitation. In Figure 3(a), the response is symmetrical for a small (<0.1 V) positive and negative excitation pulses and only a "dispersing" wave train trails

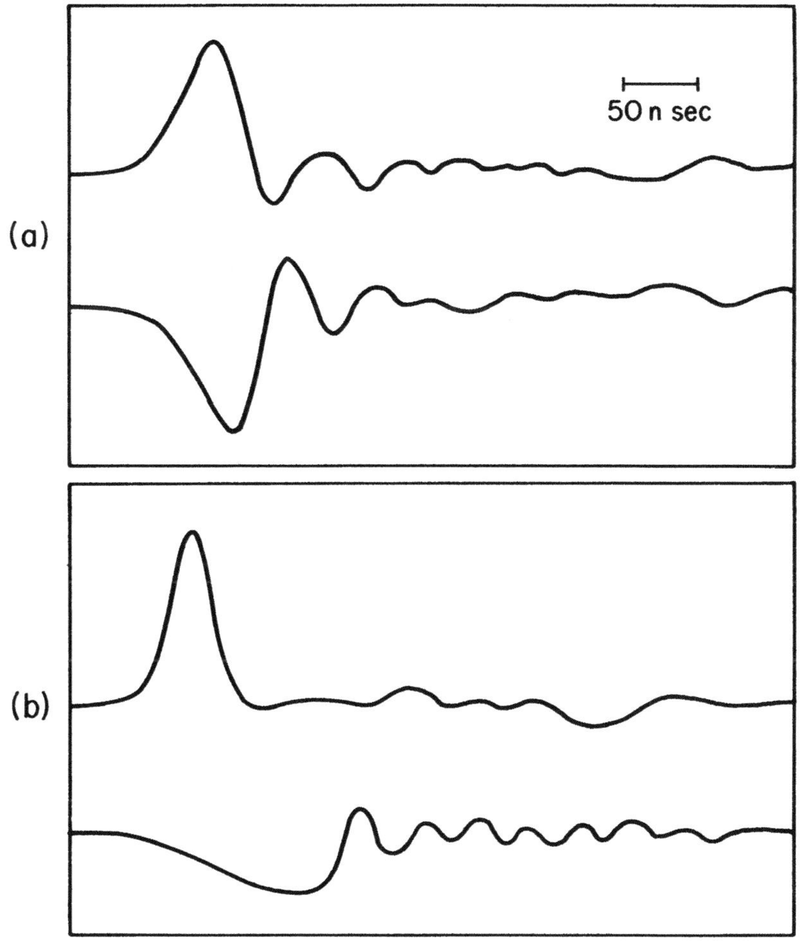

Figure 3. Response of the transmission line at a fixed distance from the point of excitation by a narrow positive or negative pulse

(a) Linear regime $V_{excitation} = \Delta V$

(b) Nonlinear regime, $V_{excitation} = 5\Delta V$

In these experiments, $|V_{excitation}| < |V_{bias}|$ and ΔV is in arbitrary units.

the first peak. Such results are expected for linear dispersive transmission lines with this particular configuration [1].

The response shown in Figure 3(b) is for an excitation pulse which is approximately 10 times larger than that in Figure 3(a). For the positive excitation pulse, the large pulse moves faster than in Figure 3(a) and the dispersing wave train is very small. For the negative excitation pulse, the dispersing wave train is observed and the large pulse moves slower than the linear case.

III. NONLINEAR WAVE EQUATION

To examine the solitary wave solutions of (2), it is convenient to re-write (5) using the experimentally observed dependence $C_N(V) = C_{NO}\left(\frac{V}{\bar{V}}\right)^{-n}$ and in dimensionless variables

$$\Gamma = \frac{V}{\bar{V}}, \quad \zeta = \sqrt{C_{NO}/C_S}\, x, \qquad \tau = \frac{1}{\sqrt{LC_S}}\, t = \omega_o t ,$$

$$M = \frac{a}{\omega_o/\sqrt{C_{NO}/C_S}}, \quad \text{and} \quad \xi = \zeta \pm M\tau .$$

Note that this equation permits both right and left traveling wave solutions. In addition, we apply the solitary wave condition that Γ, Γ', and $\Gamma'' \to 0$ as $|\xi| \to \infty$ which sets the constant $\alpha - \beta/aL = 0$. Equation (5) can then be written as

(8)
$$M^2 \frac{d^2\Gamma}{d\xi^2} + \Gamma - M^2\Gamma^{1-n} = 0 .$$

A first integral of (8) is

(9)
$$\frac{M^2}{2}\left(\frac{d\Gamma}{d\xi}\right)^2 + \frac{\Gamma^2}{2} - M^2\,\frac{\Gamma^{2-n}}{2-n} = 0 ,$$

where the constant of integration is set equal to zero since for a soliton, both Γ and $\frac{d\Gamma}{d\xi} \to 0$ as $|\xi| \to \infty$. We now integrate (9) from $\xi = 0$ where the pulse height is maximum, say Γ_1, to $\xi = \xi$ where $\Gamma = \Gamma$. (This integration is facilitated if one lets $\Gamma = \gamma^{1/n}$.) We find

$$(10)\qquad \Gamma = \begin{cases} \left[\dfrac{M^2}{2-n}\left(1 + \cos\dfrac{n\xi}{M}\right)\right]^{1/n}, & \left|\dfrac{n\xi}{M}\right| \le \pi\,, \\ 0, & \left|\dfrac{n\xi}{M}\right| > \pi\,. \end{cases}$$

Equation (10) predicts the shape of the soliton. We find that the Mach number M for a particular applied pulse can be computed in terms of the applied pulse at $x = 0$, $t = 0$ $(\xi = 0)$. We find

$$M = \left(\frac{2-n}{2}\right)^{1/2} \Gamma_{app}^{n/2}\,.$$

We finally address ourselves to the problem of specifying the limits on n. This can be accomplished by using the condition that Γ, $\frac{d\Gamma}{d\xi}$, and $\frac{d^2\Gamma}{d\xi^2}$ must be continuous. We find that $0 < n < 1$ which includes the experimental value of $n = 1/3$.

IV. EXPERIMENTS

Several experiments have been performed to illustrate some properties unique to those solitary waves which are "solitons" [12]. As an example, we illustrate the "Recurrence Phenomena" in Figure 4.

By recurrence, we mean that a signal will undergo nonlinear and dispersive distortions as it propagates and will closely approximate its original form at some later position L (the classic Fermi-Pasta-Ulam problem [13]). This distance L can be calculated in some cases and the calculation clearly demonstrates the interaction and competition between nonlinear and dispersive effects. For a sine-wave excitation of frequency ω, this would involve the generation of harmonics $n\omega$ by the nonlinear element. However, signals at $n\omega$ and nk may not satisfy the dispersion relation for the media and would, therefore, not propagate. They will, however, beat with a signal that does actually satisfy the dispersion relation for the media at a frequency ω^* and nk. At some distance L, the original signal can be recovered. By making ω an appreciable

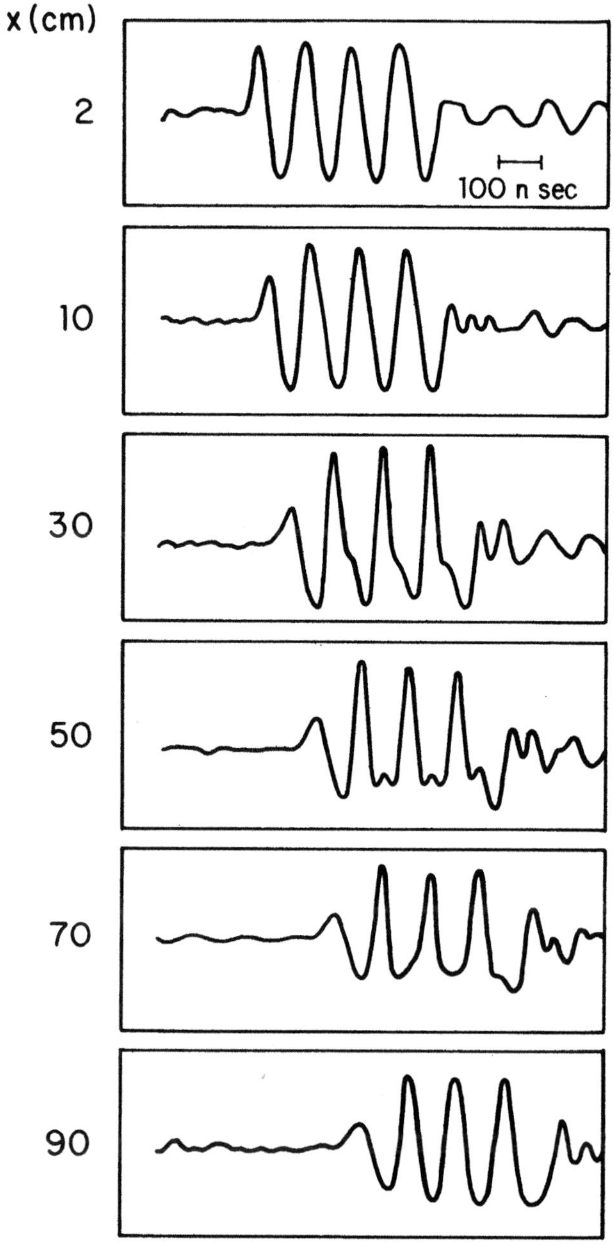

Figure 4. Experimental observation of the Fermi-Pasta-Ulam recurrence phenomena at various points on the transmission line.

fraction of ω_o, say $\omega \sim \omega_o/2$, we need only examine the beating between two signals by following the calculations of Tappert and Judice [14] and Ikezi [15] who examined this phenomenon for ion acoustic waves in plasmas. They found that $L \sim \omega^{-3}$.

This phenomenon, first observed experimentally by Hirota and Suzuki [6], is shown in Figure 4. The use of a sine-wave burst allows us to separate any reflected signals from the transmitted ones. We would identify the recurrence length as being 90 cm at this frequency. Using a spectrum analyzer, we found the second-harmonic signal changed such that it was a minimum at $x \sim 0$ and L and a maximum at $x \sim L/2$. Similar experiments were performed using different values of exciting frequency and the predicted dependence of $L \sim \omega^{-3}$ was confirmed.

In Figure 5, we examine the response at various points on the line when a "ramp" excitation is applied. Note the initial steepening of the wavefront

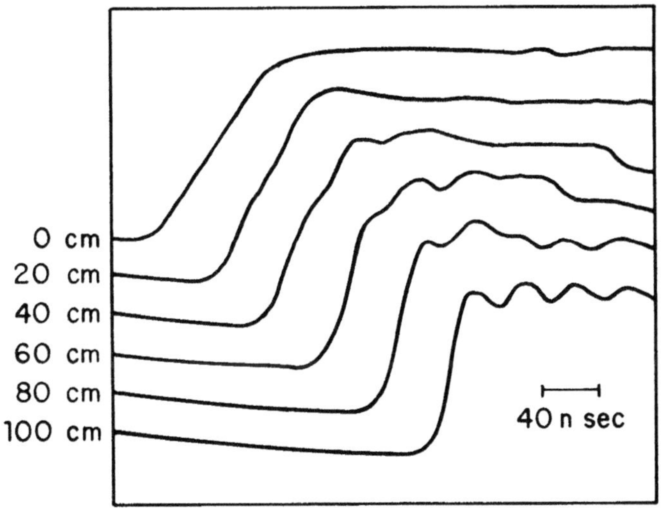

Figure 5. Experimental observation of the formation of a shock.

which is reminiscent of the formation of a shock and is due to the dominance of the nonlinear effects over the dispersive effects. For a true steady state

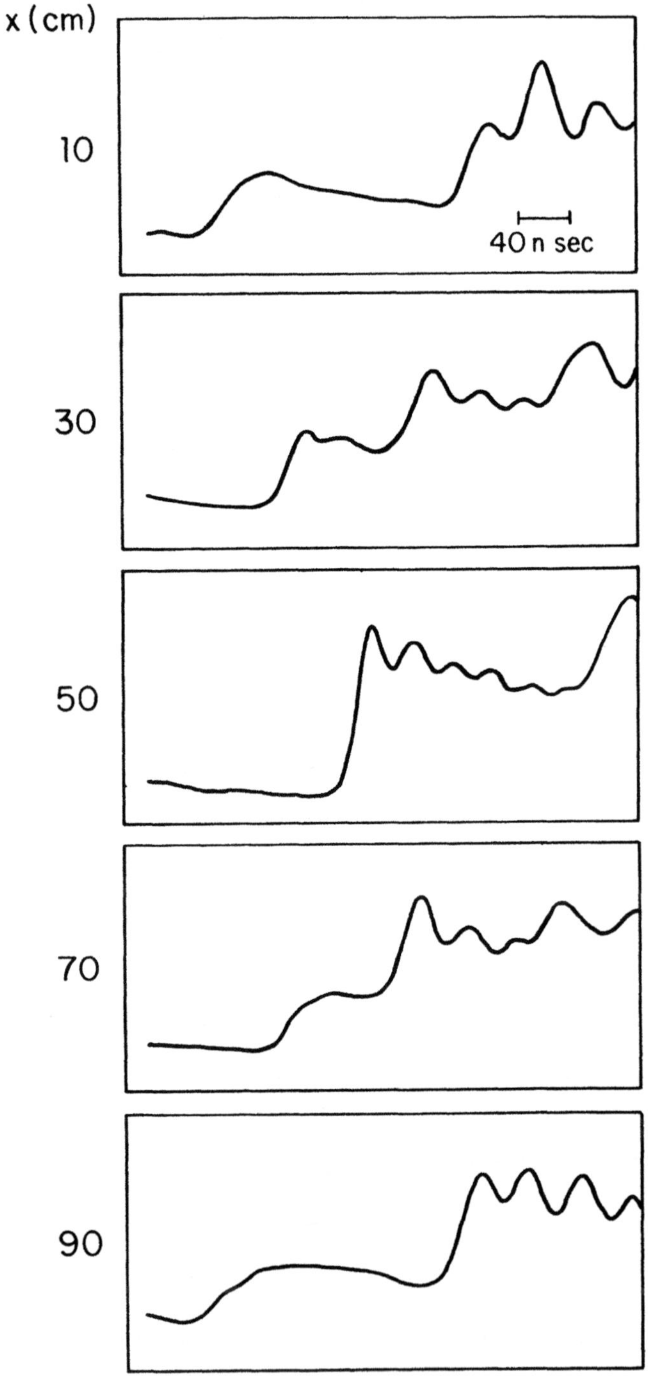

Figure 6. Experimental observation of the "collision" of two shocks.

shock to exist, we must have some loss mechanism present (see e.g., Montgomery and Joyce [16]). As none exists, we might expect that this "shock" will eventually break up into a train of solitary waves. This conjecture seems to be borne out when we observe the "nondestructive" nature of the collision of two "shocks", one launched from each end of the transmission line, as shown in Figure 6. See the paper by Scott, Chu, and McLaughlin [10] for further comments on the collision of two solitons.

V. CONCLUSION

In this paper, we have reviewed several properties of solitary waves that were ascertained from experiments on a nonlinear dispersive transmission line.

ACKNOWLEDGMENT

The author wishes to acknowledge his collaborators on this study, H. Hsuan, W. F. Ames, J. A. Kolosick, D. L. Landt, and C. M. Burde. In addition, the author wishes to acknowledge R. M. Miura for his interest and stimulating comments.

REFERENCES

[1] D.L. LANDT, C.M. BURDE, H.C.S. HSUAN, AND K.E. LONNGREN, An experimental simulation of waves in plasma, Amer. J. Phys. 40 (1972), 1493.

[2] J. KOLOSICK, D.L. LANDT, H.C.S. HSUAN, AND K.E. LONNGREN, Experimental study of solitary waves in a nonlinear transmission line, Appl. Phys. 2 (1973), 129.

[3] J. KOLOSICK, D.L. LANDT, H.C.S. HSUAN, AND K.E. LONNGREN, Properties of solitary waves as observed on a nonlinear dispersive transmission line, Proc. IEEE 62 (1974), 578.

[4] K.E. LONNGREN, D.L. LANDT, C.M. BURDE, AND J.A. KOLOSICK, Observation of shocks on a nonlinear dispersive transmission line, IEEE Trans. Circuits and Systems, CAS-22 (1975), 376.

[5] K.E. LONNGREN, H.C.S. HSUAN, AND W.F. AMES, On the soliton, invariant and shock solutions of a fourth-order nonlinear equation, J. Math. Anal. Appl., 52 (1975), 538-545.

[6] R. HIROTA AND K. SUZUKI, Studies on lattice solitons by using electrical networks, J. Phys. Soc. Japan 28 (1970), 1366.

[7] R. HIROTA AND K. SUZUKI, Theoretical and experimental studies of lattice solitons in nonlinear lumped networks, Proc. IEEE 61 (1973), 1483.

[8] K.A. GORSHKOV, L.A. OSTROVSKII, V.V. PAPKO, AND E.N. PELINOVSKII, Solitary electromagnetic waves and parametric generation of pulses in nonlinear wave systems, Proc. Intl. Symp. on Electromagnetic Wave Theory (Tbilisi, USSR), Science Press, Moscow, USSR (1971), pp. 139-149.

[9] L.A. OSTROVSKII, V.V. PAPKO, AND E.N. PELINOVSKII, Solitary electromagnetic waves in nonlinear lines, Radiophys. and Quantum Electronics 15 (1974), 438. [Russian original: Izv. Vysš. Učebn. Zaved. Radiofizika 15 (1972), 580.]

[10] A.C. SCOTT, F.Y.F. CHU, AND D.W. MCLAUGHLIN, The soliton: a new concept in applied science, Proc. IEEE 61 (1973), 1443.

[11] K.E. LONNGREN, H.C.S. HSUAN, D.L. LANDT, C.M. BURDE, G. JOYCE, I. ALEXEFF, W.D. JONES, H.J. DOUCET, A. HIROSE, H. IKEZI, S. AKSORNKITTI, M. WIDNER, AND K. ESTABROOK, Properties of plasma waves defined by the dispersion relation $D(k,\omega) = 1 - \omega_o^2/\omega^2 + k_o^2/k^2 = 0$, IEEE Trans. Plasma Science PS-2 (1974), 93.

[12] N.J. ZABUSKY AND M.D. KRUSKAL, Interaction of solitons in a collisionless plasma and the recurrence of initial states, Phys. Rev. Lett 15 (1965), 240.

[13] E. FERMI, J.R. PASTA, AND S.M. ULAM, Studies of nonlinear problems in Collected Works of Enrico Fermi, Vol. II, Univ. of Chicago Press, Chicago Ill., 1965, p. 478.

[14] F.D. TAPPERT AND C.N. JUDICE, Recurrence of nonlinear ion acoustic waves, Phys. Rev. Lett. 29 (1972), 1308.

[15] H. IKEZI, Experiments on ion acoustic solitary waves, Phys. Fluids 16 (1973), 1668.

[16] D. MONTGOMERY AND G. JOYCE, Shock-like solutions of the electrostatic Vlasov equation, J. Plasma Phys. 3 (1969), 1.

STIMULATED RAMAN AND BRILLOUIN SCATTERING AND THE INVERSE METHOD*

Flora Ying Fun Chu†

Electrical and Computer Engineering Department
University of Wisconsin
Madison, Wisconsin 53706

I. INTRODUCTION

Stimulated Raman and Brillouin spectroscopy is used as a tool for studying the vibrational energy levels of molecules and of certain atomic groups in crystals and liquids. A laser beam at frequency ω_1 irradiates the Raman (Brillouin) medium of length L. A spectral analysis of the scattered wave will yield the frequency ω_2, which is shifted from ω_1 by an integer multiple of ω_3, the vibrational frequency of the molecule. A sketch of this experimental procedure is shown in Figure 1.

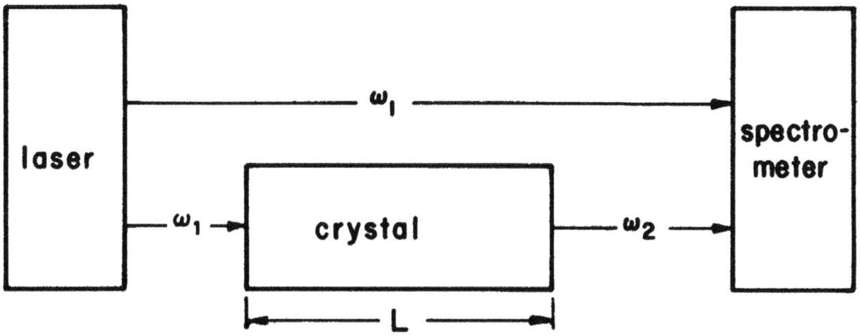

Figure 1. Sketch of experimental process used to detect stimulated Raman and Brillouin scattering.

*Supported by National Science Foundation under Grant No. GK-37552.

†Present Address: Department of Electrical Engineering, Massachusetts Institute of Technology, Cambridge, Massachusetts, 02139.

Stimulated Raman scattering (SRS) and stimulated Brillouin scattering (SBS) [1] are the scattering of coherent light by the vibrational levels of an atom or molecule. SRS is the scattering of light by optical phonons while SBS is the scattering of light by acoustical phonons. These processes can be viewed as parametric processes whereby the incident light wave at frequency ω_1 produces a coupling between the scattered wave at frequency ω_2 and the vibrational level of the molecules in the Raman (Brillouin) medium. Two types of scattering processes can occur. If the molecules in the medium are initially unexcited, the incident photon will be absorbed while simultaneously a phonon at ω_3 will propagate into the medium and the scattered photon (called the Stokes photon in SRS) at ω_2, where

(1) $$\omega_1 = \omega_2 + \omega_3 \ ,$$

will be emitted. If the molecules are initially excited, the incident photon at ω_1 and a phonon at ω_3 are absorbed while the scattered photon (called the anti-Stokes photon in SRS) at ω_2, where

(2) $$\omega_1 + \omega_3 = \omega_2 \ ,$$

will be emitted. This scattered emission depends on the molecules initially being excited to $\hbar\omega_3$ above the ground energy level where $\hbar = h/2\pi$ (h = Planck's constant). At any temperature T_o, if n_o is the population density of the molecules in the ground state, the population density in the excited state is $n_o e^{-\hbar\omega_3/k_B T_o}$ where k_B is the Boltzmann constant. The intensity of the scattering which obeys (2) is therefore a factor of $e^{-\hbar\omega_3/k_B T_o}$ lower than the intensity of the scattering which obeys (1). Scattered light is also emitted at frequencies $\omega_2 = \omega_1 \pm n\omega_3$, where $n = 2, 3, 4, \ldots$ and $\omega_2 > 0$. However, the intensities of these scatterings are much lower since they are higher order processes involving the simultaneous absorption or emission of two or more phonons at frequency ω_3.

Lamb [2] has studied the self-induced transparency phenomenon (SIT) (a phenomenon whereby ultrashort light pulses can propagate through an atomic medium as if it were transparent) by studying the interaction between a two-level atomic system and an electromagnetic wave. He has shown that the equations describing this phenomenon can be solved exactly by the inverse scattering method [3], [4]. The inverse scattering method has been shown to solve a large number of nonlinear partial differential equations. It is a technique whereby an initial-value problem of a nonlinear partial differential equation can be solved exactly through a series of linear techniques. It has been put in the following elegant form by Lax [5]. Consider a nonlinear partial differential equation $\phi_t = N(\phi)$ where N denotes a nonlinear operator on some suitable space of functions. If there exist linear operators L and B, which are functions of ϕ such that L obeys the eigenvalue equation

(3) $$L\psi = \lambda\psi \quad \text{(L equation)}$$

and B determines the time evolution of the wave function ψ by the equation

(4) $$i\psi_t = B\psi \quad \text{(B equation)},$$

then λ, the eigenvalue in (3) is independent of time if L and B satisfy the operator equation $iL_t = BL - LB$ when ϕ satisfies the nonlinear equation $\phi_t = N(\phi)$. If such operators can be found, the solution of $\phi_t = N(\phi)$ reduces to the solution of (3) and (4) [6].

Following the suggestion of Steudal [7], we treat SRS and SBS as the dual of the self-induced transparency problem (SIT) by viewing them as the interaction of a system of molecules with two electromagnetic waves instead of the interaction of an electromagnetic wave with a two level system. Equations similar to those of the SIT problems are obtained in Section II and the corresponding L and B equations are shown to be easily derived in Section III. The SRS and SBS problem can thus be solved exactly. In Section IV, we show that the L equations for SRS are actually the equations which describe the incident and scattered electric field amplitudes. Solitons and breather solutions for SRS are obtained in Section V.

II. DERIVATION OF EQUATIONS FOR STIMULATED RAMAN AND BRILLOUIN SCATTERING [7]

To develop the equations for SRS and SBS, it is assumed that the medium is initially unexcited and all scattering of electromagnetic waves occur only at frequency ω_2 where

(1) $$\omega_2 = \omega_1 - \omega_3 \ .$$

Also, the Raman (Brillouin) medium is infinitely long, i.e. in Figure 1, $L \to \infty$. We will first derive the equations for SRS. Following Yariv [1], the Raman medium is assumed to consist of harmonic oscillators, each oscillator representing one molecule. Since SRS is the scattering of light by the optical phonons which have zero group velocity, the harmonic oscillators are assumed to be independent of each other giving them zero group velocity in the laboratory frame. If X is the normal vibrational coordinate of a molecule, the equation of motion for a harmonic oscillator is

(5) $$X_{tt} + \omega_R^2 X = aE^2 - qE \ ,$$

where ω_R is the resonant vibrational frequency, aE^2 (a = constant) is the nonlinear interaction between the molecules and the electric field E, and q is the electronic charge of the molecule. The one-dimensional wave equation for the propagation of the electric field is

(6) $$E_{xx} - \frac{1}{u^2} E_{tt} = 2\mu_o a(XE)_{tt} \ ,$$

where u is the group velocity of the electromagnetic wave, μ_o is the permeability of the medium, and $2aXE$ is the nonlinear polarization. E and X are assumed to have the form

(7a) $$E = \frac{1}{2}\left(\sum_{j=1}^{2} \mathcal{E}_j(x,t) e^{i(\omega_j t - \kappa_j x + \gamma_j)} + c.c. \right) \ ,$$

(7b) $$X = \frac{1}{2}\left(\chi e^{i(\omega_3 t - \kappa_3 x + \gamma_3)} + \text{c.c.} \right) ,$$

where the E_j's and χ are slowly varying functions of space and time and the γ_j's are constants. In (7a), E_1 and E_2 are the amplitudes of the incident and scattered electric fields, respectively.

We assume that (i) the amplitude of the waves are large only at the frequencies ω_1, ω_2, and ω_3, (ii) ω_1/κ_1 and ω_2/κ_2, the phase velocities of the incident and Stokes waves, are equal to u, the group velocity of the electric field, (iii) $\kappa_1-\kappa_2-\kappa_3 = \Delta\kappa$, $\gamma_1-\gamma_2-\gamma_3 = 0$, and (iv) a is a small quantity. Then balancing the coefficients of $e^{i\omega_j t}$, $j = 1,2,3$ and keeping only first-order terms, (5) becomes

(8a) $$\chi_t - i\delta\chi = -iq_3 E_1 E_2^* e^{-i\Delta\kappa x}$$

and (6) becomes

(8b) $$\frac{1}{u} E_{1t} + E_{1x} = -iq_1 E_2 \chi e^{i\Delta\kappa x} ,$$

(8c) $$\frac{1}{u} E_{2t} + E_{2x} = -iq_2 E_1 \chi^* e^{-i\Delta\kappa x} ,$$

where $\delta \equiv \dfrac{(\Delta\omega)^2 - 2(\Delta\omega)\omega_3}{2\omega_3}$, $q_3 = \dfrac{a}{2\omega_3}$, $q_j = \dfrac{a\mu_o}{2\kappa_j}\left(\omega_j - 2\omega_3(\Delta\omega) + (\Delta\omega)^2\right)$,

$j = 1,2$ and $\Delta\omega \equiv \omega_3 - \omega_R$ defines the difference between the frequency of vibration of the oscillators ω_3 and the resonant frequency of the oscillator ω_R [8].

Equation (8), which describe the Raman medium, incident electric and scattered electric fields become

(9a) $$Y_\tau = i\delta Y - iA_1 A_2^* e^{-i\Delta\kappa\zeta} ,$$

(9b) $$A_{1\zeta} = - iA_2 Y e^{i\Delta\kappa\zeta} ,$$

(9c) $$A_{2\zeta} = - iA_1 Y^* e^{-i\Delta\kappa\zeta} ,$$

under transformation to coordinates moving with the group velocity of the electric field, i.e.

$$\zeta = x, \ \tau = t - x/u \ ,$$

and normalization of the dependent variables,

$$A_1 = \sqrt{q_2 q_3}\, \mathcal{E}_1 \ , \quad A_2 = \sqrt{q_1 q_3}\, \mathcal{E}_2 \ , \quad Y = \sqrt{q_1 q_2}\, \chi \ .$$

Writing

$$\mathcal{A}_1 = A_1 e^{-i\Delta\kappa\zeta} \ , \quad \mathcal{A}_2 = A_2 e^{i\Delta\kappa\zeta} \ , \quad \mathcal{Y} = Y e^{-i\Delta\kappa\zeta} \ ,$$

(9) become

(10a) $$\mathcal{Y}_\tau = i\delta\mathcal{Y} - i\mathcal{A}_1\mathcal{A}_2^* \ ,$$

(10b) $$\mathcal{A}_{1\zeta} = - i\Delta\kappa\mathcal{A}_1 - i\mathcal{Y}\mathcal{A}_2 \ ,$$

(10c) $$\mathcal{A}_{2\zeta} = i\Delta\kappa\mathcal{A}_2 - i\mathcal{Y}^*\mathcal{A}_1 \ .$$

Following [9] and defining the quantities

(11a) $$\mathcal{U} = i\mathcal{A}_1\mathcal{A}_2^* \ ,$$

(11b) $$\mathcal{W} = \mathcal{A}_1\mathcal{A}_1^* - \mathcal{A}_2\mathcal{A}_2^* \ ,$$

(10) can be put in a form similar to the SIT equations obtained by Lamb [2,3]. This enables us to obtain easily the equations needed for the inverse scattering method. In (11b), $\mathcal{W}$ defines the difference in intensity between the incident and Stokes waves. In terms of these new variables, (10) are

(12a) $$\mathcal{Y}_\tau = i\delta\mathcal{Y} - \mathcal{U} \ ,$$

(12b) $$\mathcal{U}_\zeta = -\mathcal{Y}\mathcal{W} - 2i\Delta\kappa\mathcal{U} \ ,$$

(12c) $$\mathcal{W}_\zeta = 2(\mathcal{U}\mathcal{Y}^* + \mathcal{U}^*\mathcal{Y}) \ .$$

When $\delta = 0$ these equations are similar to those which describe SIT.

SBS [1] is the scattering of light by acoustic waves in a crystal.

Unlike the optical phonons in SRS, these acoustical vibrations can support a wave with nonzero group velocity. However, SBS can be described by the same set of normalized equations (12). This can be seen by examining the equation of motion for $\bar{X}$, the deviation of a point in the fluid or crystal from equilibrium [1],

$$T\bar{X}_{xx} - \rho\bar{X}_{tt} = -\bar{a}(E^2)_x \tag{13}$$

and the one-dimensional wave equation for the electric field E,

$$E_{xx} - \frac{1}{u^2}E_{tt} = 2\mu_o\bar{a}(E\bar{X}_x)_{tt} \tag{14}$$

where T and ρ are elastic constants, $\bar{a}$ is constant, $-\bar{a}(E^2)_x$ is the net electrostrictive force, and $2\bar{a}E\bar{X}_x$ is the nonlinear polarization of the Brillouin medium. If E and $\bar{X}$ are assumed to have the form

$$E = \frac{1}{2}\left[\sum_{j=1}^{2} \mathcal{E}_j e^{i(\omega_j t-\kappa_j x+\gamma_j)} + \text{c.c.}\right] ,$$

$$\bar{X} = \frac{1}{2}\left[\tilde{\chi} e^{i(\omega_3 t-\kappa_3 x+\gamma_3)} + \text{c.c.}\right] ,$$

and $\bar{a}$ is small, as before, then the coefficients of $e^{i\omega_j t}$ in (13) and (14) can be balanced to give

$$Y_\tau = i\bar{\delta}\, Y - iA_1A_2^* e^{-i\Delta\kappa(\zeta+\tau)} , \tag{15a}$$

$$A_{1\zeta} = -\, iYA_2 e^{i\Delta\kappa(\zeta+\tau)} , \tag{15b}$$

$$A_{2\zeta} = -\, iY^* A_1 e^{-i\Delta\kappa(\zeta+\tau)} , \tag{15c}$$

where

$$\zeta = \frac{u}{v-u}(-x+vt) , \quad \tau = \frac{v}{v-u}(x-ut) ,$$

with $v = \sqrt{T/\rho}$, and the dependent variables are normalized to

$$A_1 = \sqrt{\bar{q}_2\bar{q}_3}\,\mathcal{E}_1,\ A_2 = \sqrt{\bar{q}_1\bar{q}_3}\,\mathcal{E}_2,\ Y = \sqrt{\bar{q}_1\bar{q}_2}\,\tilde{\chi},$$

with $\bar{q}_j = \dfrac{-i\mu_o \bar{a}\kappa_3}{2\kappa_j}\omega_j^2$, j=1,2, $\bar{q}_3 = \dfrac{i\bar{a}}{2T\kappa_3}(\kappa_2-\kappa_1)$, and $\bar{\delta} = \dfrac{1}{2\kappa_3}\left(\kappa_3^2 - \dfrac{\omega_3^2}{v}\right)$.

Equations (15) are analogous to (9), therefore with similar transformations on (15) one finds that (12) also describe SBS. In the rest of this paper, we will discuss results with respect to SRS but similar conclusions can be drawn for SBS.

III. THE INVERSE SCATTERING METHOD EQUATIONS FOR SRS

Since the SRS equations (12) are similar to the equations which describe SIT, following Lamb [3] and Ablowitz, Kaup, and Newell [4], we find the L equations to be

$$\psi_{1\zeta} + i\lambda\psi_1 = \mathcal{V}\psi_2 ,$$

(16)

$$\psi_{2\zeta} - i\lambda\psi_2 = -\mathcal{V}^*\psi_1 ,$$

and the B equations are

$$\psi_{1\tau} = \frac{i}{4}\left(\frac{\mathcal{W}}{\lambda-\Delta\kappa} + \frac{i}{2}\delta\right)\psi_1 + \frac{i}{2}\frac{\mathcal{U}}{\lambda-\Delta\kappa}\psi_2 ,$$

(17)

$$\psi_{2\tau} = \frac{i}{2}\frac{\mathcal{U}^*}{\lambda-\Delta\kappa}\psi_1 - \frac{i}{4}\left(\frac{\mathcal{W}}{\lambda-\Delta\kappa} + \frac{i}{2}\delta\right)\psi_2 .$$

IV. RELATIONSHIP OF THE INVERSE SCATTERING METHOD EQUATIONS TO THE EQUATIONS WHICH DESCRIBE STIMULATED RAMAN SCATTERING

To date, most of the L and B equations associated with nonlinear equations have been discovered by guessing their general form. Ablowitz, Kaup, Newell, and Segur [10] studied a set of L and B equations and from them determined the nonlinear p.d.e.'s associated with them. However, given a nonlinear p.d.e., there is still no systematic procedure to find the corresponding L and B equations. Lamb [3] studied the SIT phenomena and obtained the L equation associated with the nonlinear partial differential equations by studying the quantum mechanics of SIT. Motivated by this, McLaughlin and Corones [11] studied the quantum mechanics associated with the propagation of magnetic flux along a Josephson transmission line and again, they found that the L equation associated with this system can be obtained from the equations which describe this phenomenon. This result is also true for SRS. The equations which describe the interaction of the electric field amplitudes with the vibrations of the molecules are equivalent to the L equations (16). To see this, we consider the equations (10b,c) which describe the propagation of the electric fields in the Raman medium. Recall that the A_j's measure the slowly varying amplitude of the incident and scattered electric fields and $\mathcal{Y}$ is a measure of the slowly varying amplitudes of the vibrational coordinates. Rewrite (10b,c) as

$$A_{1\zeta} + i\Delta\kappa A_1 = \mathcal{Y}(-iA_2) \ ,$$

(18)

$$(-iA_2)_\zeta - i\Delta\kappa(-iA_2) = -\mathcal{Y}^*(A_1) \ ,$$

and comparing (18) with (16), it is seen that these two sets of equations are equivalent if

$$A_1 \equiv \psi_1 \ , \ -iA_2 \equiv \psi_2 \ , \quad \text{and} \quad \Delta\kappa = \lambda \ .$$

Thus, the equations (18) describing the SRS phenomenon are equivalent to the L equations (16) of SRS when λ, the eigenvalue is equal to $\Delta\kappa$, the value of λ which makes the B equations of SRS (17) singular.

V. SOLITON SOLUTIONS FOR STIMULATED RAMAN SCATTERING

Once the L and B equations for SRS are found, the solution $\mathcal{Y}(\zeta,\tau)$ can be found. From Ablowitz, Kaup, Newell, and Segur [12] it is given by

(19) $$\mathcal{Y}(\zeta,\tau) = -2K(\zeta,\zeta,\tau)$$

where $K(\zeta,y,\tau)$ obeys the Marchenko equation

(20a) $$K(\zeta,y,\tau) = F(\zeta+y,\tau) - \int_{-\infty}^{\zeta}\int_{-\infty}^{\zeta} F(y+s_1,\tau)F^*(s_1+s_2,\tau)K(\zeta,s_2,\tau)ds_1 ds_2$$

with

(20b) $$F(y,\tau) = \frac{1}{2\pi}\int_{-\infty}^{\infty} \frac{b(\lambda,\tau)}{a(\lambda,\tau)} e^{-i\lambda y}\, d\lambda - i \sum_{j=1}^{N} \bar{c}_j(\lambda_j,\tau) e^{-i\lambda_j y}$$

where N is the number of discrete eigenvalues for (16). If f and g are Jost functions which satisfy (16) with boundary conditions

$$\lim_{x\to-\infty} f\, e^{i\lambda x} = \begin{pmatrix}1\\0\end{pmatrix},$$

$$\lambda \text{ real},$$

$$\lim_{x\to+\infty} g\, e^{-i\lambda x} = \begin{pmatrix}0\\1\end{pmatrix},$$

and for λ complex, $\bar{g} \equiv \begin{pmatrix} g_2^*(\zeta,\lambda) \\ -g_1(\zeta,\lambda^*) \end{pmatrix}$, then a and b are defined by

$$f = bg + a\bar{g}$$

when λ is real. $a(\lambda,\tau)$ and $b(\lambda,\tau)$ can be analytically continued to the upper half of the λ plane and $\bar{c}$ is defined as

$$\bar{c} \equiv \frac{b^*(\lambda^*,\tau)}{a^*(\lambda^*,\tau)} .$$

The time dependence of a, b, and $\bar{c}$ are calculated to be

(21a) $$b(\lambda,\tau) = b(\lambda,0) e^{-\frac{i}{2}\left(\frac{1}{\lambda-\Delta\kappa} + 2\delta\right)\tau},$$

(21b) $$a(\lambda,\tau) = a(\lambda,0) ,$$

and

(21c) $$\bar{c}(\lambda_j,\tau) = \bar{c}(\lambda_j,0) e^{\frac{i}{2}\left(\frac{1}{\lambda-\Delta\kappa} + 2\delta\right)\tau}$$

$$= \bar{c}(\lambda_j,0) e^{\omega_{rj}\tau + i\omega_{ij}\tau} ,$$

where

$$\omega_{rj} \equiv \frac{\lambda_{ij}}{2[(\lambda_{rj}-\Delta\kappa)^2 + \lambda_{ij}^2]} ,$$

(21d) $\lambda_j \equiv \lambda_{rj}+\lambda_{ij}$

$$\omega_{ij} \equiv \frac{\lambda_{rj}-\Delta\kappa}{2[(\lambda_{rj}-\Delta\kappa)^2 + \lambda_{ij}^2]} + \delta .$$

The values $b(\lambda,0)$, $a(\lambda,0)$, and $\bar{c}(\lambda_j,0)$ are determined by the initial conditions.

If $b(\lambda,0) = 0$, and $N = 1$ in (19) and (20), the solution $\mathcal{Y}$ is of the form

(22) $\mathcal{Y}(\zeta,\tau) =$

$$\frac{2i\bar{c}_1(\lambda_1,0)}{|\bar{c}_1(\lambda_1,0)|} \lambda_{i1} e^{i(\omega_{i1}\tau-2\lambda_{r1}\zeta)} \operatorname{sech}\left(2\lambda_{i1}\zeta+\omega_{r1}\tau + \ln \frac{|\bar{c}_1(\lambda_1,0)|}{2\lambda_{i1}}\right) .$$

This is a one-soliton solution [12]. If $b(\lambda,0) = 0$, and there are N discrete eigenvalues, the solution $\mathcal{Y}$ will consist of N such solitons, these solitons will interact nonlinearly, but, asymptotically $\mathcal{Y}$ will consist of the superposition of N solitons in the form (22). From Zakharov and Shabat [12], the formula for $|\mathcal{Y}|^2$ when $\mathcal{Y}$ is an N-soliton solution is

$$|\mathcal{Y}|^2 = 4 \ln[\det(1+ZZ^*)]_{xx}$$

where Z is an $N\times N$ matrix with elements defined by

$$Z_{j\ell} = \frac{\sqrt{\bar{c}_j\bar{c}_\ell^*}}{(\lambda_j-\lambda_\ell^*)} e^{-i(\lambda_j-\lambda_\ell^*)\zeta} .$$

One interesting type of solution is the "breather". These are real solutions of (16) formed by two solitons whose associated eigenvalues λ_1 and λ_2 are related by

$$\lambda_1 = -\lambda_2^* \equiv \lambda \equiv \lambda_r + i\lambda_i .$$

Using (19) and (20), a breather solution is

$$Y = \frac{-2\lambda_i}{|c(\lambda,0)|\lambda_r}\left(\frac{\lambda_r\cosh\theta_1\sin\theta_2 + \lambda_i\sinh\theta_1\cosh\theta_2}{\cosh^2\theta_1 + \frac{\lambda_i^2}{\lambda_r^2}\cos^2\theta_2}\right)$$

where

$$\theta_1 = 2\lambda_i\zeta + \omega_r\tau + \ln\left(\frac{|c(\lambda,0)|}{2\lambda_i}\frac{\lambda_r}{|\lambda|}\right)$$

$$\theta_2 = -2\lambda_r\zeta + \omega_i\tau - i\ln\left(\frac{c(\lambda,0)}{|c(\lambda,0)|}\frac{|\lambda|}{\lambda}\right)$$

with $\bar{c}_1 = -\bar{c}_2^* \equiv c$ and $\omega_r = \omega_{r1}$, $\omega_i = \omega_{i1}$ where ω_{r1} and ω_{i1} are defined by (21d). A sketch of $|Y|$ when $\Delta\kappa = 0.1$, $\delta = -0.1$, $\lambda_r = \lambda_i = 1$, $c = \lambda\sqrt{2}$ is shown in Figure 2.

When $b(\lambda,0) \neq 0$, (20) cannot be solved exactly. However, the contribution from the continuous spectrum decays away and only the discrete eigenvalues contribute to the solution Y. Therefore, asymptotically, the solution will consist only of solitons.

VI. CONCLUSIONS

The equations describing SRS and SBS are solved under the following assumptions:

(i) The Raman and Brillouin media are lossless. The L and B equations have only been found when the conductivity of the medium and the energy dissipation of the vibrational wave are zero.

(ii) The electric fields and normal vibrations vary only in one dimension.

(iii) The Raman and Brillouin media are infinite. Obviously, in an SRS or SBS experiment, the medium cannot be infinite. If the length of the medium is L (see Figure 1), the solutions Y are valid only when the width w of a soliton is much less than L. If w is taken to be the half width

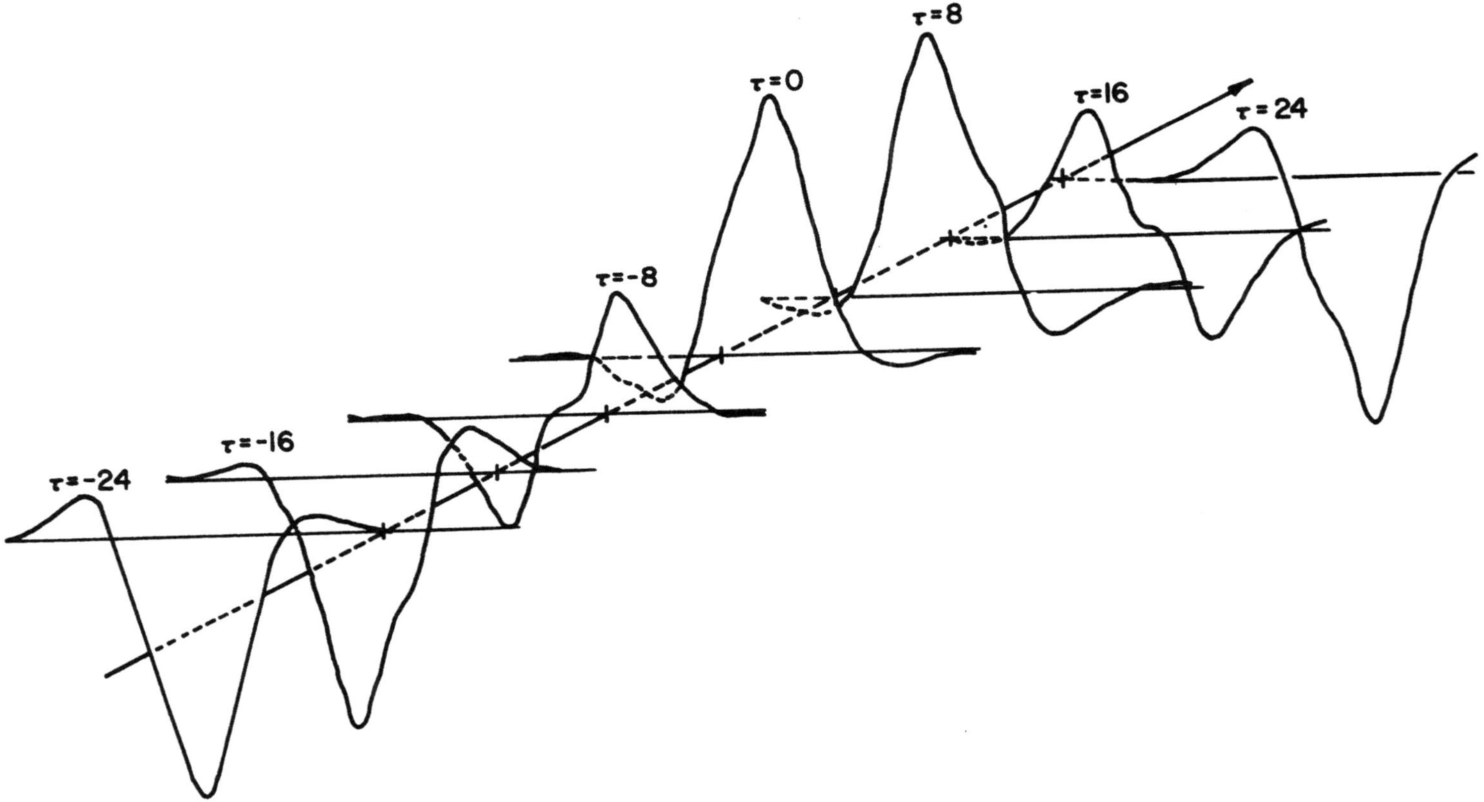

Figure 2. Sketch of a breather solution y vs. horizontal axes ζ at $\tau = -24, -16, -8, 0, 8, 16, 24$ for $\Delta\kappa = 0.1$, $\delta = -0.1$, $\lambda_r = \lambda_i = 1$, $c = \sqrt{2}\lambda$.

of a single soliton, then the equations are valid only if

$$w = 2.64(2\lambda_i - \frac{\omega_r}{u})^{-1} << L$$

where $\lambda = \lambda_r + i\lambda_i$ depends on the initial conditions. On the other hand, the size of a soliton cannot be too small. The molecules in the Raman medium are assumed to be harmonic oscillators and a soliton has to be much wider than the spacing between the molecules. Taking the average distance between the molecules to be approximately 10^{-8}m, the equations describing SRS are valid if the initial conditions are such that

$$10^{-8}\text{m} << 2.64(2\lambda_i - \frac{\omega_r}{u})^{-1} << L \, .$$

(iv) The function $b(\lambda,0) = 0$. If $b(\lambda,0) \neq 0$, the Marchenko equation (20a) cannot be solved exactly. Analytic solutions can only be obtained asymptotically, when the contribution from the continuous spectrum has decayed. Therefore, the length of the Raman medium L has to be much longer than the break-up distance of the solitons.

Thus, under the above assumptions, the inverse scattering method enables one to obtain explicit solutions for SRS. However, there are still some shortcomings of the method in its present stage. The inverse scattering operators have been found in the (ζ,τ)-coordinates, i.e. coordinates which are moving with velocity u, the group velocity of the electromagnetic field. This implies that if an initial-value problem is to be done the initial condition

$$V(\zeta,\tau=0) = V(x,t=x/u)$$

has to be specified. This is not a realistic initial condition. The realistic initial-value problem with initial condition $V(x,t=0) = V(\zeta,\tau=-\zeta/u)$ cannot be solved unless this initial condition is modified or some new method for utilizing the inverse scattering method for the SRS equations is found.

ACKNOWLEDGMENT

I would like to thank Professor A. C. Scott and P. Rissman for helpful discussions and T. Thousand for computational assistance.

REFERENCES

[1] A. YARIV, Quantum Electronics, John Wiley & Sons, New York, N. Y., 1967, Chaps. 23 and 25.

[2] G.L. LAMB, JR., Analytical descriptions of ultrashort optical pulse propagation in a resonant medium, Rev. Modern Phys. 43 (1971), 99-124.

[3] G.L. LAMB, JR., Phase variation in coherent-optical pulse propagation, Phys. Rev. Lett. 31 (1973), 196-199.

[4] M.J. ABLOWITZ, D.J. KAUP, AND A.C. NEWELL, Coherent pulse propagation, a dispersive, irreversible phenomenon, J. Mathematical Phys. 15 (1974), 1852-1858.

[5] P.D. LAX, Integrals of nonlinear equations of evolution and solitary waves, Comm. Pure Appl. Math. 21 (1968), 647-690.

[6] F.Y.F. CHU, Physical applications of the soliton theory, Ph.D. Thesis, University of Wisconsin, 1974.

[7] H. STEUDEL, Stimulierte Ramanstreuung mit ultrakurzen Lichtimpulsen, Exp. Tech. der Physik 20 (1972), 409-415.

[8] S.A. AKHMANOV, K.N. DRABOVICH, A.P. SUKHORUKOV, AND A.S. CHIRKIN, Stimulated Raman scattering in a field of ultrashort light pulses, Soviet Physics JETP 32 (1971), 266-273.

[9] R.P. FEYNMAN, F.L. VERNON, JR., AND R.W. HELLWARTH, Geometrical representation of the Schrödinger equation for solving maser problems, J. Appl. Phys. 28 (1957), 49-52.

[10] M.J. ABLOWITZ, D.J. KAUP, A.C. NEWELL, AND H. SEGUR, The inverse scattering transform-Fourier analysis for nonlinear problems, Studies in Appl. Math. 53 (1974), 249-315.

[11] D.W. MCLAUGHLIN AND J. CORONES, On semi-classical radiation theory and the inverse method, Phys. Rev. A 10 (1974), 2051-2062.

[12] A.C. SCOTT, F.Y.F. CHU, AND D.W. MCLAUGHLIN, The soliton: A new concept in applied science, Proc. IEEE 61 (1973), 1443-1483.

[13] V.E. ZAKHAROV AND A.B. SHABAT, Exact theory of two-dimensional self-focusing and one-dimensional self-modulation of waves in nonlinear media, Soviet Physics JETP 34 (1972), 62-69.

DIRECT METHOD OF FINDING EXACT SOLUTIONS OF NONLINEAR EVOLUTION EQUATIONS

Ryogo Hirota

Department of Mathematics and Physics
Ritsumeikan University
Kita-ku, Kyoto, Japan

I. INTRODUCTION

The main purpose of this article is to present a direct and systematic way of finding exact solutions of certain nonlinear evolution equations. We transform the nonlinear evolution equations into bilinear differential equations of the following special form

$$F\left(\frac{\partial}{\partial t} - \frac{\partial}{\partial t'}, \frac{\partial}{\partial x} - \frac{\partial}{\partial x'}\right)a(t,x)b(t',x')\Big|_{t=t',x=x'} = 0 ,$$

which we solve exactly using a kind of perturbational approach. Examples shown in this article are the modified Korteweg-deVries equation, the nonlinear Schrödinger equation with normal dispersion, wave-wave interactions, the two-dimensional Korteweg-deVries equations, and the two-dimensional sine-Gordon equation. In the appendices we mention the N-soliton form of the modified Korteweg-deVries equation and the Cole-Hopf transformation.

II. PRELIMINARIES

In order to illustrate the present method we first consider the modified Korteweg-deVries equation [1], [2]

$$v_t + \alpha(v^3)_x + v_{xxx} = 0 \tag{1}$$

where subscripts indicate partial differentiations.

We solve the equation by the usual perturbation method. We expand v as a power series in a parameter ϵ

$$v = \epsilon v_1 + \epsilon^3 v_3 + \epsilon^5 v_5 + \cdots . \tag{2}$$

Substituting (2) into (1) and collecting terms with the same powers of ϵ, we obtain

$$(\frac{\partial}{\partial t} + \frac{\partial^3}{\partial x^3})v_1 = 0 ,$$

(3) $$(\frac{\partial}{\partial t} + \frac{\partial^3}{\partial x^3})v_3 = - \alpha(v_1^3)_x ,$$

$$(\frac{\partial}{\partial t} + \frac{\partial^3}{\partial x^3})v_5 = - 3\alpha(v_1^2 v_3)_x ,$$

and so on. If we start with a solution

$$v_1 = \exp(\eta_1)$$

where $\eta_1 = p_1 x - \Omega_1 t$, $\Omega_1 = p_1^3$, p_1 being an arbitrary complex parameter, we would find a particular series solution of (3) with

$$v_3 = \alpha \frac{3p_1}{3\Omega_1 - (3p_1)^3} \exp(3\eta_1) ,$$

(4)

$$v_5 = 3\alpha^2 \left(\frac{5p_1}{5\Omega_1 - (5p_1)^3} \right) \left(\frac{3p_1}{3\Omega_1 - (3p_1)^3} \right) \exp(5\eta_1) ,$$

and so on, provided that all homogeneous solutions are assumed to be zero.

One may ask whether the perturbation solution converges. In the region where the real part of η_1 is negative and sufficiently large, the perturbation series converges and the solution has a physical meaning. However, in the region where the real part of η_1 is positive and large, each term in the series diverges and the solution loses its physical meaning.

We meet similar difficulties in other fields of physics. The difficulties are sometimes overcome by summing up the divergent terms in the perturbation series. In the present case, the perturbation terms are simple and we could attempt to sum up the series. We actually have

(5)
$$v_3 = -(\alpha/8p_1^2)\exp(3\eta_1) ,$$
$$v_5 = (\alpha/8p_1^2)^2\exp(5\eta_1) ,$$
$$v_7 = -(\alpha/8p_1^2)^3\exp(7\eta_1) .$$

A formal sum of the series gives

(6)
$$v = \frac{\epsilon\exp(\eta_1)}{1 + \epsilon^2(\alpha/8p_1^2)\exp(2\eta_1)} .$$

Substituting (6) into (1) we find that this is an exact solution and is equivalent to the well-known single soliton solution for real η_1

(7)
$$v = (2/\alpha)^{\frac{1}{2}}p_1/\cosh(\eta_1) .$$

If we started with a solution

$$v_1 = \exp(\eta_1) + \exp(\eta_2) ,$$

this would lead to the series for the 2-soliton solution, but it would require a considerable amount of work before we found an explicit solution.

We propose a simpler method to find an exact solution. The functional form of the solution (6) suggests that it is convenient to replace v by G/F and to find the equations to be satisfied by G and F.

We remark that our approach is similar to the Padé approximation. The (n,m) Padé approximant to a function $f(\epsilon)$ is the ratio of two polynomials, the numerator of degree m and the denominator of degree n. The Padé approximant has been widely applied to many branches of physics [3].

Substituting $v = G/F$ into (1) we find

$$(G_tF - GF_t)/F^2 + 3\alpha(G/F)^2(G_xF - GF_x)/F^2$$

(8)
$$+ (G_{xxx}F - 3G_{xx}F_x - 3G_xF_{xx} - GF_{xxx})/F^2$$

$$+ 6(FG_xF_x^2 + FGF_xF_{xx} - GF_x^3)/F^4 = 0 ,$$

which apparently has a more complicated form than the original equation for v, but it has at least one simple solution, namely

$$F = 1 + \epsilon^2(\alpha/8p_1^2)\exp(2\eta_1), \quad G = \epsilon\exp(\eta_1) .$$

Keeping this in mind we simplify (8) by decoupling it into two equations. After several trials we find (8) is transformed into

(9)
$$(G_tF - GF_t + G_{xxx}F - 3G_{xx}F_x + 3G_xF_{xx} - GF_{xxx})/F^2$$

$$+ 3(G_xF - GF_x)[\alpha G^2 - 2(FF_{xx} - F_x^2)]/F^4 = 0 .$$

Therefore, F and G will be a solution to the above equation, and hence v ($= G/F$) is a solution of (1) if they satisfy the following coupled equations

$$G_tF - GF_t + G_{xxx}F - 3G_{xx}F_x + 3G_xF_{xx} - GF_{xxx} = 0 .$$

(10)

$$2(FF_{xx} - F_x^2) = \alpha G^2 ,$$

which is expressed as

$$[(\frac{\partial}{\partial t} - \frac{\partial}{\partial t'}) + (\frac{\partial}{\partial x} - \frac{\partial}{\partial x'})^3]G(x,t)F(x',t')\Big|_{x = x', t = t'} = 0 ,$$

(11)

$$(\frac{\partial}{\partial x} - \frac{\partial}{\partial x'})^2F(x,t)F(x',t)\Big|_{x = x'} = \alpha G^2 .$$

Note that the differential operator

$$(\frac{\partial}{\partial t} - \frac{\partial}{\partial t'}) + (\frac{\partial}{\partial x} - \frac{\partial}{\partial x'})^3 ,$$

is related to the linear operator of the modified Korteweg-deVries equation

$$\frac{\partial}{\partial t} + \frac{\partial^3}{\partial x^3} \quad .$$

We solve (10) or (11) by expanding F and G as power series in a parameter ϵ

$$F = 1 + \epsilon^2 f_2 + \epsilon^4 f_4 + \cdots \; , \tag{12}$$

$$G = \epsilon g_1 + \epsilon^3 g_3 + \cdots \; .$$

Substituting (12) into (11) and collecting terms with the same power of ϵ, we have[1]

$$\left(\frac{\partial}{\partial t} + \frac{\partial^3}{\partial x^3}\right) g_1 = 0$$

$$2\frac{\partial^2}{\partial x^2} f_2 = \alpha g_1^2 \; , \tag{13}$$

$$\left(\frac{\partial}{\partial t} + \frac{\partial^3}{\partial x^3}\right) g_3 + \left[\left(\frac{\partial}{\partial t} - \frac{\partial}{\partial t'}\right) + \left(\frac{\partial}{\partial x} - \frac{\partial}{\partial x'}\right)^3\right] g_1(x,t) f_2(x',t')\Big|_{x=x',t=t'} = 0 \; ,$$

$$2\frac{\partial^2}{\partial x^2} f_4 + \left(\frac{\partial}{\partial x} - \frac{\partial}{\partial x'}\right)^2 f_2(x,t) f_2(x',t')\Big|_{x=x',t=t'} = 2\alpha g_1 g_3 \; ,$$

and so on. We start with a solution

$$g_1 = \exp(\eta_1) + \exp(\eta_2)$$

where $\eta_i = p_i x - \Omega_i t - \gamma_i$, $\Omega_i = p_i^3$, $\gamma_i =$ constant, for $i = 1,2$. Note that the above equations are linear differential equations with known inhomogeneous terms. Solving these equations successively, we find the particular solutions

$$f_2 = b(2,0)\exp(2\eta_1) + b(1,1)\exp(\eta_1+\eta_2) + b(0,2)\exp(2\eta_2) \; ,$$

$$g_3 = b(2,1)\exp(2\eta_1+\eta_2) + b(1,2)\exp(\eta_1 + 2\eta_2) \; , \tag{14}$$

$$f_4 = b(2,2)\exp(2\eta_1+2\eta_2)$$

---

[1] A similar procedure was also introduced by G. B. Whitham [4, p. 581].

where

$$b(2,0) = \alpha/2(2p_1)^2, \quad b(1,1) = \alpha/(p_1+p_2)^2, \quad b(0,2) = \alpha/2(2p_2)^2 ,$$

$$(15) \quad b(2,1) = b(2,0)(p_1-p_2)^2/(p_1+p_2)^2, \quad b(1,2) = b(0,2)(p_1-p_2)^2/(p_1+p_2)^2 ,$$

$$b(2,2) = b(2,0)b(0,2)(p_1-p_2)^4/(p_1+p_2)^4 ,$$

and all higher-order terms are zero. An exact 2-soliton solution is now written as

$$(16) \quad v = G/F$$

$$= \frac{\epsilon(e^{\eta_1}+e^{\eta_2}) + \epsilon^3(\frac{p_1-p_2}{p_1+p_2})^2 (b(2,0)e^{\eta_1} + b(0,2)e^{\eta_2})e^{\eta_1+\eta_2}}{1+\epsilon^2[b(2,0)e^{2\eta_1}+ \frac{\alpha}{(p_1+p_2)^2} e^{\eta_1+\eta_2}+b(0,2)e^{2\eta_2}]+\epsilon^4 b(2,0)b(0,2)(\frac{p_1-p_2}{p_1+p_2})^4 e^{2\eta_1+2\eta_2}} .$$

In Appendix I we describe the form of N-soliton solution of the modified Korteweg-deVries equation [1].

We summarize the procedure as follows: Transform the nonlinear evolution equation, for example by $v = G/F$, and find the special bilinear differential equations for F and G, one of which is related to the linear differential operator. Expand F and G as power series in a parameter ϵ and determine the coefficients by perturbation theory.

Before we apply the present method to other nonlinear evolution equations, we introduce the operators D_t, D_x, and various products of them by

$$(17) \quad \begin{aligned} D_t^n a\cdot b &\equiv (\frac{\partial}{\partial t} - \frac{\partial}{\partial t'})^n a(t)b(t')\Big|_{t=t'} , \\ D_x^n a\cdot b &\equiv (\frac{\partial}{\partial x} - \frac{\partial}{\partial x'})^n a(x)b(x')\Big|_{x=x'} . \end{aligned}$$

With this notation, (11) is expressed simply as

$$(D_t + D_x^3)G\cdot F = 0 \ ,$$

(18)

$$D_x^2 F\cdot F = \alpha G^2 \ .$$

We list some properties of these operators which are used later:

(I) $$D_x^m a\cdot 1 = (\frac{\partial}{\partial x})^m a \ .$$

(II) $$D_x^m a\cdot b = (-1)^m D_x^m b\cdot a \ ,$$

(II.1) $$D_x^m a\cdot a = 0 \quad \text{for odd} \quad m \ .$$

(III) $$D_x^m a\cdot b = D_x^{m-1}(a_x\cdot b - a\cdot b_x) \ ,$$

(III.1) $$D_x^m a\cdot a = 2D_x^{m-1} a_x\cdot a \quad \text{for even} \quad m \ ,$$

(III.2) $$D_x D_t a\cdot a = 2D_x a_t\cdot a$$

$$= 2D_t a_x\cdot a \ .$$

(IV) $$D_x^m \exp(\alpha_1 x)\cdot\exp(\alpha_2 x) = (\alpha_1 - \alpha_2)^m \exp[(\alpha_1 + \alpha_2)x],$$

(IV.1) $$(D_t^n + D_x^m)\exp(\alpha_1 x + \beta_1 t)\cdot\exp(\alpha_2 x + \beta_2 t)$$

$$= \left\{ \frac{(\beta_1-\beta_2)^n + (\alpha_1-\alpha_2)^m}{(\beta_1+\beta_2)^n + (\alpha_1+\alpha_2)^m} \right\} (D_t^n + D_x^m)\exp[(\alpha_1+\alpha_2)x + (\beta_1+\beta_2)t]\cdot 1 \ .$$

(V) $$\exp(\epsilon D_x)a(x)\cdot b(x) = a(x+\epsilon)b(x-\epsilon) \ .$$

(VI) $$\exp(\epsilon D_x)ab\cdot cd = [\exp(\epsilon D_x)a\cdot c][\exp(\epsilon D_x)b\cdot d] \ ,$$

$$= [\exp(\epsilon D_x)a\cdot d][\exp(\epsilon D_x)b\cdot c] \ ,$$

(VI.1) $$D_x ab\cdot c = (\frac{\partial a}{\partial x})bc + a(D_x b\cdot c) \ ,$$

(VI.2) $$D_x^2 ab\cdot c = (\frac{\partial^2 a}{\partial x^2})bc + 2(\frac{\partial a}{\partial x})D_x b\cdot c + a(D_x^2 b\cdot c) \ ,$$

(VI.3) $$D_x^m \exp(\alpha x)a\cdot\exp(\alpha x)b = \exp(2\alpha x)(D_x^m a\cdot b) .$$

(VII) $$\exp(\epsilon\frac{\partial}{\partial x})[b/a] = [\exp(\epsilon D_x)b\cdot a]/[\cosh(\epsilon D_x)a\cdot a] ,$$

(VII.1) $$\frac{\partial}{\partial x}(\frac{b}{a}) = \frac{D_x b\cdot a}{a^2} ,$$

(VII.2) $$\frac{\partial^2}{\partial x^2}(\frac{b}{a}) = \frac{D_x^2 b\cdot a}{a^2} - (\frac{b}{a})\frac{D_x^2 a\cdot a}{a^2} ,$$

(VII.3) $$\frac{\partial^3}{\partial x^3}(\frac{b}{a}) = \frac{D_x^3 b\cdot a}{a^2} - 3\frac{(D_x b\cdot a)}{a^2}\frac{D_x^2 a\cdot a}{a^2} .$$

(VIII) $$2\cosh(\epsilon\frac{\partial}{\partial x})\log f = \log[\cosh(\epsilon D_x)f\cdot f] ,$$

(VIII.1) $$\frac{\partial^2}{\partial x^2}\log f = \frac{1}{2f^2}(D_x^2 f\cdot f) ,$$

(VIII.2) $$\frac{\partial^4}{\partial x^4}\log f = \frac{1}{2f^2}(D_x^4 f\cdot f) - 6[\frac{1}{2f^2}(D_x^2 f\cdot f)]^2 .$$

All of these properties are easily verified so we show only (VII). We have

$$\exp(\epsilon\frac{\partial}{\partial x})[b(x)/a(x)] = b(x+\epsilon)/a(x+\epsilon) .$$

On the other hand, we have from (V)

$$\exp(\epsilon D_x)b(x)\cdot a(x) = b(x+\epsilon)a(x-\epsilon) ,$$

$$\cosh(\epsilon D_x)a(x)\cdot a(x) = a(x+\epsilon)a(x-\epsilon) ,$$

and hence

$$\exp(\epsilon D_x)b(x)\cdot a(x)/\cosh(\epsilon D_x)a(x)\cdot a(x) = b(x+\epsilon)/a(x+\epsilon) ,$$

which proves (VII). Equations (VII.1)-(VII.3) are obtained by expanding (VII) as a power series in ϵ and equating terms with the same power of ϵ. Other properties of the D operators are described in another paper by the author [5].

III. NONLINEAR SCHRÖDINGER EQUATION

It is known that a wide class of nonlinear wave equations can be approximated by the nonlinear Schrödinger equation [6]

(19) $$i\psi_t + \psi_{xx} + q|\psi|^2\psi = 0 \; .$$

For $q > 0$, the equation has an N-envelope-soliton solution [7], [8]. For $q < 0$, Hasegawa and Tappert [9] found a dark pulse solution (envelope-hole solution)

(20) $$|\psi|^2 = \rho_o^2[1 - a^2 \operatorname{sech}^2(\rho_o a\xi)]$$

where ρ_o is a real constant, $a^2 < 1$, and $\xi = x - vt$.

We consider envelope-hole solutions of (19) with the boundary condition

(21) $$|\psi|^2 = \rho_o^2 \quad \text{as} \quad |x| \to \infty \; .$$

Substituting $\psi = G/F$, F real, into (19) and using (VII.1) and (VII.2), we have

(22) $$\frac{iD_t G\cdot F}{F^2} + \frac{D_x^2 G\cdot F}{F^2} - \frac{G}{F}\frac{D_x^2 F\cdot F}{F^2} - 2\,\frac{GG^*}{F^2}\,\frac{G}{F} = 0 \; ,$$

where q is chosen to be -2 for convenience and G^* is the complex conjugate of G. The coupled equations for F and G are chosen as

(23) $$(iD_t + D_x^2 - \lambda)G\cdot F = 0 \; ,$$

(24) $$(D_x^2 - \lambda)F\cdot F = -2GG^* \; ,$$

where λ is a constant to be determined. We note that (24) is transformed by (VIII.1) to

(25) $$|\psi|^2 = \frac{\lambda}{2} - \frac{\partial^2}{\partial x^2}\log F \; .$$

We expand F and G as power series in ϵ

$$F = 1 + \epsilon f_1 + \epsilon^2 f_2 + \dots ,$$
(26)
$$G = g_o(1 + \epsilon g_1 + \epsilon^2 g_2 + \dots) ,$$

where $f_1, f_2, \dots, g_1, g_2, \dots$ are assumed to go to zero as $x \to -\infty$. The boundary condition (21) as $x \to -\infty$ gives $g_o g_o^* = \rho_o^2$ and asymptotically we have the equations

(27) $$(iD_t + D_x^2 - \lambda) g_o \cdot 1 = 0 ,$$

(28) $$(D_x^2 - \lambda) 1 \cdot 1 = -2 g_o g_o^* ,$$

which uniquely determine g_o and λ :

(29) $$g_o = \rho_o \exp(i\theta)$$

where

$$\theta = kx - \omega t , \quad \omega = k^2 + 2\rho_o^2, \; k = \text{ real constant},$$

and

(30) $$\lambda = 2\rho_o^2 .$$

Substituting (26) into (23) and (24), and using (VI.1), (VI.2) and (29), we have

(31) $$[i(D_t + 2kD_x) + D_x^2](1 + \epsilon g_1 + \dots)\cdot(1 + \epsilon f_1 + \dots) = 0$$

and

(32) $$(D_x^2 - 2\rho_o^2)(1 + \epsilon f_1 + \dots)\cdot(1 + \epsilon f_1 + \dots)$$
$$= -2\rho_o^2(1 + \epsilon g_1 + \dots)(1 + \epsilon g_1^* + \dots) .$$

Collecting terms with the same powers of ϵ, we have

$$[i(D_t + 2kD_x) + D_x^2](g_1 \cdot 1 + 1 \cdot f_1) = 0 \ ,$$

(33)

$$[D_x^2 - 2\rho_o^2](f_1 \cdot 1 + 1 \cdot f_1) = -2\rho_o^2(g_1 + g_1^*) \ ,$$

$$[i(D_t + 2kD_x) + D_x^2](g_2 \cdot 1 + g_1 \cdot f_1 + 1 \cdot f_2) = 0 \ ,$$

(34)

$$[D_x^2 - 2\rho_o^2](f_2 \cdot 1 + f_1 \cdot f_1 + 1 \cdot f_2) = -2\rho_o^2(g_2 + g_1 g_1^* + g_2^*) \ ,$$

$$[i(D_t + 2kD_x) + D_x^2](g_3 \cdot 1 + g_2 \cdot f_1 + g_1 \cdot f_2 + 1 \cdot f_3) = 0 \ ,$$

(35)

$$[D_x^2 - 2\rho_o^2](f_3 \cdot 1 + f_2 \cdot f_1 + f_1 \cdot f_2 + 1 \cdot f_3) = -2\rho_o^2(g_3 + g_2 g_1^* + g_1 g_2^* + g_3^*) \ ,$$

$$[i(D_t + 2kD_x) + D_x^2](g_4 \cdot 1 + g_3 \cdot f_1 + \dots + 1 \cdot f_4) = 0 \ ,$$

(36)

$$[D_x^2 - 2\rho_o^2](f_4 \cdot 1 + f_3 \cdot f_1 + \dots + 1 \cdot f_4) = -2\rho_o^2(g_4 + g_3 g_1^* + \dots + g_4^*) \ ,$$

and so on.

The envelope-hole solution of Hasegawa and Tappert is obtained by choosing the following pair of starting solutions of (33)

(37) $$f_1 = \exp(\eta), \quad g_1 = b \exp(\eta) \ ,$$

where

$$\eta = px - \Omega t \ ,$$

(38)

$$\Omega = p\left[2k \pm \sqrt{(2\rho_o)^2 - p^2}\right],$$

p being a real constant satisfying the condition $p^2 \leq (2\rho_o)^2$ and

$$b = -\frac{p^2+i(\Omega-2kp)}{p^2-i(\Omega-2kp)} .$$

Substituting the pair of starting solutions (37) into (34), we find that all higher-order terms can be taken to be zero. Accordingly, the envelope-hole solution is given by

(39) $$\psi = \rho_o \exp(i\theta)[1 + b\exp(\eta)]/[1 + \exp(\eta)] .$$

Using (25) we have

(40) $$|\psi|^2 = \rho_o^2 - (p/2)^2 \operatorname{sech}^2(\eta/2)$$

which becomes (20) by identifying $\eta = 2\rho_o a\xi$ and $p = 2\rho_o a$.

For constructing a 2-envelope-hole solution we start with the following solution pair

(41) $$f_1 = \exp(\eta_1) + \exp(\eta_2), \quad g_1 = b_1\exp(\eta_1) + b_2\exp(\eta_2) ,$$

where

(42) $$\eta_i = p_i x - \Omega_i t + \text{constant} ,$$

(43) $$\Omega_i = p_i[2k \pm \sqrt{(2\rho_o)^2-p_i^2}\ ] ,$$

p_i being a real constant satisfying the condition $p_i^2 \leq (2\rho_o)^2$ and

(44) $$b_i = \exp(2i\phi_i)$$

with

(45) $$\tan\phi_i = -p_i^2/(\Omega_i - 2kp_i)$$

Substituting (41) into (34) we find

$$f_2 = a_{12} \exp(\eta_1 + \eta_2) ,$$

(46)

$$g_2 = b_{12} \exp(\eta_1 + \eta_2) ,$$

where

(47) $$a_{12} = \left[\frac{\sin\frac{1}{2}(\phi_1 - \phi_2)}{\sin\frac{1}{2}(\phi_1 + \phi_2)}\right]^2 , \quad b_{12} = b_1 b_2 a_{12} , \quad p_i = 2\rho_o \sin\phi_i .$$

Substituting (41) and (46) into (35) and (36) we find that all higher terms can be taken to be zero. Accordingly we have an exact 2-envelope-hole solution

$$\psi = \rho_o \exp(i\theta)[g/f] ,$$

(48) $$f = 1 + \exp(\eta_1) + \exp(\eta_2) + a_{12} \exp(\eta_1 + \eta_2) ,$$

$$g = 1 + \exp(\eta_1+2i\phi_1) + \exp(\eta_2+2i\phi_2) + a_{12}\exp(\eta_1+\eta_2+2i\phi_1+2i\phi_2).$$

The form of (48) suggests that the N-envelope-hole solution* can be expressed as

$$\psi = \rho_o \exp(i\theta)[g/f] ,$$

(49) $$f = \sum_{\mu=0,1} \exp\left[\sum_{i>j}^{(N)} A_{ij}\mu_i\mu_j + \sum_{i=1}^{N} \mu_i\eta_i\right] ,$$

$$g = \sum_{\underset{\sim}{\mu}=0,1} \exp\left[\sum_{i>j}^{(N)} A_{ij}\mu_i\mu_j + \sum_{i=1}^{N} \mu_i(\eta_i + 2i\phi_i)\right] ,$$

where

$$\exp(A_{ij}) = \left[\frac{\sin\frac{1}{2}(\phi_i - \phi_j)}{\sin\frac{1}{2}(\phi_i + \phi_j)}\right]^2 .$$

$$\eta_i = p_i x - \Omega_i t$$

$$p_i = 2\rho_o \sin\phi_i$$

$$\Omega_i = 2\rho_o \sin\phi_i[2k - 2\rho_o \cos\phi_i] ,$$

---

*The author is very grateful to Dr. M. Wadati for pointing out to him that an N-envelope-hole solution of the nonlinear Schrödinger equation has been reported by Zakharov and Shabat [10].

all ϕ_i are distinct real constants, $\sum_{\underset{\sim}{\mu}=0,1}$ is the summation over all possible combinations of $\mu_1 = 0,1$, $\mu_2 = 0,1$, ... , $\mu_N = 0,1$, and $\sum_{i>j}^{(N)}$ indicates the summation over all possible pairs chosen from N elements. The conjecture (49) can be proved by the same procedure used in reference [8].

IV. WAVE-WAVE INTERACTIONS

1. Two-Wave Interaction

We consider the following equations [11]

$$(50)\qquad \begin{aligned} &(\frac{\partial}{\partial t} + v_1 \frac{\partial}{\partial x})\phi_1 = - \phi_1\phi_2 , \\ &(\frac{\partial}{\partial t} + v_2 \frac{\partial}{\partial x})\phi_2 = \phi_1\phi_2 , \end{aligned}$$

which describe the interaction of two waves ϕ_1 and ϕ_2 propagating with velocities v_1 and v_2, respectively [12]. Let $\phi_1 = G_1/F$, $\phi_2 = G_2/F$, then we have

$$(51)\qquad \begin{aligned} &D_1 G_1 \cdot F = - G_1 G_2 , \\ &D_2 G_2 \cdot F = G_1 G_2 , \end{aligned}$$

where

$$D_i = D_t + v_i D_x, \quad \text{for } i = 1,2 .$$

We expand F, G_1, and G_2 as power series in ϵ

$$(52)\qquad \begin{aligned} F &= 1 + \epsilon f_1 + \epsilon^2 f_2 + \dots , \\ G_1 &= \epsilon g_1 + \epsilon^2 g_2 + \dots , \\ G_2 &= \epsilon h_1 + \epsilon^2 h_2 + \dots . \end{aligned}$$

Substituting (52) into (51) and collecting terms with the same power in ϵ, we have

$$D_1 g_1 \cdot 1 = 0 \ ,$$

(53)

$$D_2 h_1 \cdot 1 = 0 \ ,$$

$$D_1(g_1 \cdot f_1 + g_2 \cdot 1) = - g_1 h_1 \ ,$$

(54)

$$D_2(h_1 \cdot f_1 + h_2 \cdot 1) = g_1 h_1 \ ,$$

and so on. From (53) we have

$$g_1 = g_1(x - v_1 t) \ ,$$

(55)

$$h_1 = h_1(x - v_2 t) \ ,$$

where $g_1(x)$ and $h_1(x)$ are arbitrary functions of x. Substituting (55) into (54), we find that all higher terms can be chosen to be zero if f_1 satisfies the following equations

$$\left(\frac{\partial}{\partial t} + v_1 \frac{\partial}{\partial x}\right) f_1 = h_1 \ ,$$

(56)

$$\left(\frac{\partial}{\partial t} + v_2 \frac{\partial}{\partial x}\right) f_1 = -g_1 \ .$$

A general solution of (56) is given by

(57)
$$f_1 = F_1(x - v_1 t) + F_2(x - v_2 t) \ ,$$

where F_1 and F_2 are related to g_1 and h_1 by

$$\left(\frac{\partial}{\partial t} + v_1 \frac{\partial}{\partial x}\right) F_2(x - v_2 t) = h_1(x - v_2 t) \ ,$$

(58)

$$\left(\frac{\partial}{\partial t} + v_2 \frac{\partial}{\partial x}\right) F_1(x - v_1 t) = -g_1(x - v_1 t) \ .$$

Accordingly we have an exact solution of (50)

$$\phi_1 = \frac{-(\frac{\partial}{\partial t} + v_2 \frac{\partial}{\partial x})F_1(x - v_1 t)}{F_1(x - v_1 t) + F_2(x - v_2 t)} ,$$

(59)

$$\phi_2 = \frac{(\frac{\partial}{\partial t} + v_1 \frac{\partial}{\partial x})F_2(x - v_2 t)}{F_1(x - v_1 t) + F_2(x - v_2 t)} ,$$

where we write $F_1 + F_2$ in place of $1 + F_1 + F_2$ because of the arbitrariness of F_1 and F_2. This form of the solution was obtained by Hashimoto [13].

The form of (59) reminds us of the Cole-Hopf transformation of Burgers equation

$$u_t - u_{xx} + 2uu_x = 0 .$$

In Appendix II we describe how the present method gives the Cole-Hopf transformation [14].

2. Volterra's Equation

Volterra's equation which describes the growth of conflicting populations in biological sciences is written as [15]

$$\frac{dN_1}{dt} = N_1(a_1 - N_2) ,$$

(60)

$$\frac{dN_2}{dt} = - N_2(a_2 - N_1) ,$$

where N_1 and N_2 are the populations of two species, prey and predator, and a_1 and a_2 are the growing and damping constants of the two species, respectively.

We note that (60) is transformed to [16], [17]

$$\frac{d}{dt} \log (1 + V/a_2) = - I ,$$

(61)

$$\frac{d}{dt} \log (1 + I/a_1) = V ,$$

by the use of

$$a_2 + V = N_1 ,$$

(62)

$$a_1 + I = N_2 .$$

Equation (61) describes a nonlinear LC circuit with the nonlinear capacitance $C(V)$ and the inductance $L(I)$ given by

(63) $$C(V) = (1/V)\log(1 + V/a_2) ,$$

(64) $$L(I) = (1/I)\log(1 + I/a_1) ,$$

where V and I are the voltage and the current in the circuit. It is known that the nonlinear LC circuit equation (61) has an oscillatory solution, whose angular frequency ω_o is approximately given by $\omega_o^2 = I/L(I)C(V)$. Accordingly the Volterra equation has an oscillatory solution.

For large a_1, (64) gives a linear inductance $L = 1/a_1$, and the ladder-type nonlinear LC network [18] constructed with this L and $C(V)$ becomes a circuit equivalent to the one-dimensional Toda lattice [19]. We have observed solitons propagating through this nonlinear network [18]. Daikoku et al [17] have shown that the ladder-type nonlinear LC circuit with the nonlinearities given by (63) and (64) also supports solitons.

We modify the Volterra equation by introducing propagation velocities v_1 and v_2 for the two species

$$\left(\frac{\partial}{\partial t} + v_1 \frac{\partial}{\partial x}\right)N_1 = N_1(a_1 - N_2) ,$$

(65)

$$\left(\frac{\partial}{\partial t} + v_2 \frac{\partial}{\partial x}\right)N_2 = - N_2(a_2 - N_1) .$$

Let $N_1 = \phi_1\exp(a_1t)$ and $N_2 = \phi_2\exp(-a_2t)$, then (65) become

$$(\frac{\partial}{\partial t} + v_1 \frac{\partial}{\partial x})\phi_1 = - \phi_1\phi_2 \exp(-a_2 t) \ ,$$

(66)

$$(\frac{\partial}{\partial t} + v_2 \frac{\partial}{\partial x})\phi_2 = \phi_1\phi_2 \exp(a_1 t) \ .$$

Equation (66) reduces to (50) if $a_1 = a_2 = 0$. We solve (66) following the same procedure used in solving (50). Let $\phi_1 = G_1/F$, $\phi_2 = G_2/F$, then we have

$$D_1 G_1 \cdot F = - G_1 G_2 \exp(-a_2 t) \ ,$$

(67)

$$D_2 G_2 \cdot F = G_1 G_2 \exp(a_1 t) \ .$$

We expand F, G_1, and G_2 as power series in ϵ

$$F = 1 + \epsilon f_1 + \epsilon^2 f_2 + \dots \ ,$$

(68)

$$G_1 = \epsilon g_1 + \epsilon^2 g_2 + \dots \ ,$$

$$G_2 = \epsilon h_1 + \epsilon^2 h_2 + \dots \ .$$

Substituting (68) into (67) and collecting terms with the same power in ϵ, we have

$$D_1 g_1 \cdot 1 = 0 \ ,$$

(69)

$$D_2 h_1 \cdot 1 = 0 \ ,$$

$$D_1(g_1 \cdot f_1 + g_2 \cdot 1) = - g_1 h_1 \exp(-a_2 t) \ ,$$

(70)

$$D_2(h_1 \cdot f_1 + h_2 \cdot 1) = g_1 h_1 \exp(a_1 t) \ ,$$

and so on. From (69) we have

$$g_1 = g_1(x - v_1 t) \ ,$$

(71)

$$h_1 = h_1(x - v_2 t) \ .$$

Substituting (71) into (70), we find that all higher terms can be chosen to be zero if f_1 satisfies

$$\left(\frac{\partial}{\partial t} + v_1 \frac{\partial}{\partial x}\right) f_1 = h_1 \exp(-a_2 t) ,$$

(72)

$$\left(\frac{\partial}{\partial t} + v_2 \frac{\partial}{\partial x}\right) f_1 = - g_1 \exp(a_1 t) .$$

An exact solution is obtained if we start with a solution pair

$$g_1 = g_o \exp[p(x - v_1 t)] ,$$

(73)

$$h_1 = h_o \exp[p(x - v_2 t)] ,$$

where g_o, h_o, and p are constants to be determined. Substituting (73) into (72) we find

(74) $$f_1 = f_o \exp[P(x - Vt)] ,$$

where

$$P = (a_1 + a_2)/(v_1 - v_2) ,$$

$$V = (a_1 v_2 + a_2 v_1)/(a_1 + a_2) ,$$

(75)

$$g_o = a_2 f_o ,$$

$$h_o = a_1 f_o .$$

Accordingly we have

$$N_1 = a_2 f_o \exp[P(x - Vt)]/[1 + f_o \exp[P(x - Vt)]] ,$$

(76)

$$N_2 = a_1 f_o \exp[P(x - Vt)]/[1 + f_o \exp[P(x - Vt)]] ,$$

which states that the boundary region of thickness $1/P$ connecting two stationary states ($N_1 = N_2 = 0$ and $N_1 = a_2$, $N_2 = a_1$) are moving with the velocity V.

3. Three-Wave Interactions

We consider the following equations for real ϕ_1, ϕ_2, and ϕ_3

$$(77)\qquad \begin{aligned} (\frac{\partial}{\partial t} + v_1 \frac{\partial}{\partial x})\phi_1 &= q_1\phi_2\phi_3 , \\ (\frac{\partial}{\partial t} + v_2 \frac{\partial}{\partial x})\phi_2 &= q_2\phi_3\phi_1 , \\ (\frac{\partial}{\partial t} + v_3 \frac{\partial}{\partial x})\phi_3 &= q_3\phi_1\phi_2 , \end{aligned}$$

where $q_1 = \pm 1$, $q_2 = \pm 1$, and $q_3 = \pm 1$. Equation (77) describes the interaction of three waves ϕ_1, ϕ_2, and ϕ_3 propagating with velocities v_1, v_2, and v_3, respectively. Zakharov and Manakov [20] solved (77) by using the inverse scattering method. We show that the exact solution of (77) can also be found by the present method.

Transforming ϕ_i by $\phi_i = G_i/F$ for $i = 1,2,3$ and substituting into (77), we have

$$(78)\qquad \begin{aligned} D_1G_1\cdot F &= q_1G_2G_3 , \\ D_2G_2\cdot F &= q_2G_3G_1 , \\ D_3G_3\cdot F &= q_3G_1G_2 , \end{aligned}$$

where

$$D_i = D_t + v_iD_x .$$

Expanding F and G_i as power series in ϵ

$$(79)\qquad \begin{aligned} F &= 1 + \epsilon^2 f_2 + \epsilon^4 f_4 + \dots , \\ G_1 &= g_{10} + \epsilon^2 g_{12} + \epsilon^4 g_{14} + \dots , \\ G_2 &= \epsilon g_{21} + \epsilon^3 g_{23} + \dots , \\ G_3 &= \epsilon g_{31} + \epsilon^3 g_{33} + \dots , \end{aligned}$$

substituting them into (78), and following the same procedure used in the previous sections, we finally obtain the exact solution of (77):

(a) For the case $g_{10} = 0$, we have

(80)
$$\phi_1 = \frac{a_{22}\exp(\eta_2 + \eta_3)}{1 + a_{02}\exp(2\eta_2) + b_{02}\exp(2\eta_3)} ,$$
$$\phi_2 = \frac{a_{21}\exp(\eta_2)}{1 + a_{02}\exp(2\eta_2) + b_{02}\exp(2\eta_3)} ,$$
$$\phi_3 = \frac{a_{31}\exp(\eta_3)}{1 + a_{02}\exp(2\eta_2) + b_{02}\exp(2\eta_3)} ,$$

where

(81)
$$\eta_2 = p_2(x - v_2t) ,$$
$$\eta_3 = p_3(x - v_3t) ,$$
$$a_{22} = q_1 \frac{a_{21}a_{31}}{2(v_1 - v_2)p_2} ,$$
$$a_{02} = q_1q_3 \frac{(a_{21})^2}{4(v_1 - v_2)(v_2 - v_3)p_2^2} ,$$
$$b_{02} = - q_1q_2 \frac{(a_{31})^2}{4(v_1 - v_3)(v_2 - v_3)p_3^2} ,$$

a_{21} and a_{31} are arbitrary constants, and p_2 and p_3 are parameters satisfying the condition

(82)
$$(v_1 - v_2)p_2 = (v_1 - v_3)p_3 .$$

The form of (80) is the same as the one obtained by Zakharov and Manakov [20]. When $v_1 > v_2 > v_3 > 0$ and $q_1 < 0$, $q_2 > 0$, $q_3 < 0$, the solution states that one pulse moving with velocity v_3 is amplified by the other pulse moving with velocity v_2.

(b) For the case $g_{10} \neq 0$, we have

(83)
$$\phi_1 = g_{10} \frac{1 - \exp(2\eta_1)}{1 + \exp(2\eta_1)} ,$$
$$\phi_2 = \frac{a_1 \exp(\eta_1)}{1 + \exp(2\eta_1)} ,$$
$$\phi_3 = \frac{b_1 \exp(\eta_1)}{1 + \exp(2\eta_1)} ,$$

where

(84)
$$\eta_1 = p_1 x - \Omega_1 t ,$$
$$(\Omega_1 - v_2 p_1)(\Omega_1 - v_3 p_1) = q_2 q_3 g_{10}^2 ,$$

and

(85)
$$a_1^2 = - q_1 q_3 4(\Omega_1 - v_1 p_1)(\Omega_1 - v_3 p_1) ,$$
$$b_1^2 = - q_1 q_2 4(\Omega_1 - v_1 p_1)(\Omega_1 - v_2 p_1) .$$

The form of (83) was obtained by Nozaki and Taniuti [21].

V. TWO-DIMENSIONAL NONLINEAR WAVE EQUATIONS

1. Two-Dimensional Korteweg-deVries Equation

We have the two dimensional Korteweg-deVries equation

(86)
$$w_{tx} + 6(ww_x)_x + w_{xxxx} \pm w_{yy} = 0$$

which is known to be associated with an inverse scattering problem [22]. Very recently Satsuma [23] found an exact three-(possibly N-)soliton solution to (86). Equation (86) reduces to

(87)
$$(D_t D_x + D_x^4 \pm D_y^2) f \cdot f = 0 ,$$

by the transformation

(88) $$w = 2 \frac{\partial^2}{\partial x^2} \log f ,$$

and by use of the boundary condition that w and x-derivatives of w go to zero as $|x| \to \infty$.

We find that the following form of f is a solution to (87)

(89) $$\begin{aligned} f = 1 &+ \exp(\eta_1) + \exp(\eta_2) + \exp(\eta_3) \\ &+ \exp(A_{12} + \eta_1 + \eta_2) + \exp(A_{13} + \eta_1 + \eta_3) + \exp(A_{23} + \eta_2 + \eta_3) \\ &+ \exp(A_{12} + A_{13} + A_{23} + \eta_1 + \eta_2 + \eta_3) , \end{aligned}$$

where

(90) $$\eta_i = p_i x + q_i y - \Omega_i t ,$$

(91) $$p_i \Omega_i - p_i^4 \mp q_i^2 = 0 ,$$

and

(92) $$\exp(A_{ij}) = - \frac{(p_i - p_j)(\Omega_i - \Omega_j) - (p_i - p_j)^4 \mp (q_i - q_j)^2}{(p_i + p_j)(\Omega_i + \Omega_j) - (p_i + p_j)^4 \mp (q_i + q_j)^2}$$
$$= \frac{3(p_i - p_j)^2 \mp [(q_i/p_i) - (q_j/p_j)]^2}{3(p_i + p_j)^2 \mp [(q_i/p_i) - (q_j/p_j)]^2} ,$$

for $i, j = 1,2,3$ where p_i and q_i are arbitrary constants.

The form of (89) suggests that the N-soliton solution of the two-dimensional Korteweg-deVries equation (86) has the same functional form as the one-dimensional case [24]:

(93) $$w = 2 \frac{\partial^2}{\partial x^2} \log f ,$$
$$f = \sum_{\underset{\sim}{\mu}=0,1} \exp\left(\sum_{i>j}^{(N)} A_{ij} \mu_i \mu_j + \sum_{i=1}^{N} \eta_i \mu_i \right) .$$

2. Two-Dimensional Sine-Gordon Equation

We have the two-dimensional sine-Gordon equation

$$\phi_{xx} + \phi_{yy} - \phi_{tt} = \sin\phi \tag{94}$$

which describes the motion of the magnetic flux quanta on a Josephson junction transmission line [25]. Equation (94) reduces to

$$\begin{aligned} &(D_x^2 + D_y^2 - D_t^2)g\cdot f = gf , \\ &(D_x^2 + D_y^2 - D_t^2)(f\cdot f - g\cdot g) = 0 , \end{aligned} \tag{95}$$

by the transformation

$$\begin{aligned} \phi &= 4 \tan^{-1}(g/f) \\ &= \text{Im } 4 \log(f + ig) , \end{aligned} \tag{96}$$

where Im indicates the imaginary part.

We have found that the following form of $f + ig$ is a three-soliton solution to (95) [26]:

$$f + ig = \sum_{\underset{\sim}{\mu}=0,1} \exp\{\sum_{i>j}^{(3)} A_{ij}\mu_i\mu_j + \sum_{i=1}^{3} (\eta_i + i\pi/2)\mu_i\} \tag{97}$$

where

$$\eta_i = p_i x + q_i y - \Omega_i t , \tag{98}$$

$$p_i^2 + q_i^2 - \Omega_i^2 = 1 , \tag{99}$$

$$\begin{aligned} \exp(A_{ij}) &= - \frac{(p_i-p_j)^2 + (q_i-q_j)^2 - (\Omega_i-\Omega_j)^2}{(p_i+p_j)^2 + (q_i+q_j)^2 - (\Omega_i+\Omega_j)^2} \\ &= - \frac{1 - d_{ij}}{1 + d_{ij}} , \end{aligned} \tag{100}$$

(101) $$d_{ij} = p_i p_j + q_i q_j - \Omega_i \Omega_j , \quad \text{for } i,j = 1,2,3,$$

and the parameters p_i, q_i, and Ω_i satisfy (99) and the condition

(102) $$\det\left(d_{ij}\right) = 0 ,$$

which is transformed to

(103) $$\det \begin{pmatrix} p_1 & p_2 & p_3 \\ q_1 & q_2 & q_3 \\ \Omega_1 & \Omega_2 & \Omega_3 \end{pmatrix} = 0 .$$

The condition (103) indicates that the N-soliton solution for $N > 3$ is very difficult to find.

APPENDIX I. N-SOLITON SOLUTION OF THE MODIFIED KDV EQUATION

N-soliton solutions of the modified Korteweg-deVries equation are derived as follows. The form of (16) suggests that F and G may be expressed as [8]

(A.1) $$F = f^2 + g^2 ,$$

(A.2) $$G = (8/\alpha)^{\frac{1}{2}} D_x g \cdot f ,$$

where

$$f = 1 - \left(\frac{p_1 - p_2}{p_1 + p_2}\right)^2 \exp(\xi_1 + \xi_2) ,$$

$$g = \exp(\xi_1) + \exp(\xi_2) ,$$

$$\exp(\xi_1) = \epsilon(\alpha/2)^{\frac{1}{2}}(1/2p_1)\exp(\eta_1) ,$$

$$\exp(\xi_2) = \epsilon(\alpha/2)^{\frac{1}{2}}(1/2p_2)\exp(\eta_2) .$$

Hence we have

$$v = (8/\alpha)^{\frac{1}{2}} \frac{\partial}{\partial x} \tan^{-1}(g/f) \tag{A.3}$$

$$= (8/\alpha)^{\frac{1}{2}} \operatorname{Im} \frac{\partial}{\partial x} \log(f + ig) .$$

Substituting (A.3) into (1), we find that v is a solution of (1) if f and g satisfy the following equations

$$(D_t + D_x^3) g \cdot f = 0 ,$$

(A.4)

$$D_x^2 (f \cdot f + g \cdot g) = 0 .$$

An expression for the N-soliton solution was found by solving these equations [1]

$$v = (8/\alpha)^{\frac{1}{2}} \operatorname{Im} \frac{\partial}{\partial x} \log(f + ig) ,$$

(A.5)

$$f + ig = \sum_{\underset{\sim}{\mu}=0,1} \exp\left(\sum_{i>j}^{(N)} A_{ij}\mu_i\mu_j + \sum_{i=1}^{N} \mu_i(\xi_i + i\pi/2) \right),$$

where

$$\xi_i = p_i x - \Omega_i t , \tag{A.6}$$

$$\Omega_i = p_i^3 ,$$

$$\exp(A_{ij}) = (p_i - p_j)^2/(p_i + p_j)^2 . \tag{A.7}$$

We note that $f + ig$ can be expressed as a determinant of a matrix M

$$f + ig = \det\left(M_{ij}\right) \tag{A.8}$$

where

$$M_{ij} = \delta_{ij} + \frac{(2p_i p_j)^{\frac{1}{2}}}{p_i + p_j} \exp[\tfrac{1}{2}(\xi_i + \xi_j + i\pi)] . \tag{A.9}$$

If we replace $\xi_i + i\pi/2$ by ξ_i and write $f + g = F$, we find that

(A.10)
$$u = 2 \frac{\partial^2}{\partial x^2} \log F$$

is the N-soliton solution of the Korteweg-deVries equation [24]

(A.11)
$$u_t + 6uu_x + u_{xxx} = 0 .$$

On the other hand, using (10) we have

(A.12)
$$v^2 = G^2/F^2$$
$$= \frac{1}{\alpha} \frac{\partial^2}{\partial x^2} \log(F^2) .$$

Therefore we have

(A.13)
$$v^2 = \frac{1}{\alpha} \frac{\partial^2}{\partial x^2} \log[\det(MM^*)] .$$

This form of the N-soliton solution was obtained by Wadati [2] with the use of the inverse scattering method.

APPENDIX II. THE COLE-HOPF TRANSFORMATION OF THE BURGERS EQUATION

The Burgers equation

(A.14)
$$u_t - u_{xx} + 2uu_x = 0$$

is transformed into the linear differential equation

(A.15)
$$f_t - f_{xx} = 0$$

by the Cole-Hopf transformation [14]

(A.16)
$$u = -(\log f)x .$$

We show how the Cole-Hopf transformation is obtained by the present method. Let $u = g/f$, we have

(A.17)
$$(D_t - D_x^2)g \cdot f = 0 ,$$

(A.18) $$D_x^2 f\cdot f = -2D_x g\cdot f \ .$$

With the help of (III.1) we find that $g = -f_x$ is a solution of (A.18). Substituting this into (A.17) and using (III.1) and (III.2), we find

$$(D_t - D_x^2) f_x\cdot f = 2D_x(f_t - f_{xx})\cdot f \ .$$

Accordingly we have that $u = -f_x/f$ is a solution of (A.14) provided that f satisfies the linear differential equation

$$f_t - f_{xx} = \lambda f$$

where λ is an arbitrary constant.

REFERENCES

[1] R. HIROTA, Exact solution of the modified Korteweg-deVries equation for multiple collisions of solitons, J. Phys. Soc. Japan 33 (1972), 1456-1458.

[2] M. WADATI, The modified Korteweg-deVries equation, J. Phys. Soc. Japan 34 (1973), 1289-1296.

[3] P.R. GRAVES-MORRIS, Ed., Pade Approximants and Their Applications, Academic Press, New York, N.Y., 1973.

[4] G.B. WHITHAM, Linear and Nonlinear Waves, John Wiley and Sons, New York, N.Y., 1974.

[5] R. HIROTA, A new form of Bäcklund transformation and its relation to the inverse scattering problem, Progr. Theoret Phys. 52 (1974), 1498-1512.

[6] T. TANIUTI AND N. YAJIMA, Perturbation method for a nonlinear wave modulation. I., J. Mathematical Phys. 10 (1969), 1369-1372.

[7] V.E. ZAKHAROV AND A.B. SHABAT, Exact theory of two-dimensional self-focusing and one-dimensional self-modulation of waves in nonlinear media, Soviet Physics JETP 34 (1972), 62-69.

[8] R. HIROTA, Exact envelope-soliton solutions of a nonlinear wave equation, J. Mathematical Phys. 14 (1973), 805-809.

[9] A. HASEGAWA AND F. TAPPERT, Transmission of stationary nonlinear optical pulses in dispersive dielectric fibers. II. Normal dispersion, Appl. Phys. Lett. 23 (1973), 171-172.

[10] V.E. ZAKHAROV AND A.B. SHABAT, Interaction between solitons in a stable medium, Soviet Physics JETP 37 (1973), 823-828.

[11] A. YOSHIKAWA AND M. YAMAGUTI, On some further properties to a certain semi-linear system of partial differential equations, Publ. Res. Inst. Math. Sci., Kyoto Univ. 9 (1974), 577-595.

[12] A. HASEGAWA, Propagation of wave intensity shocks in nonlinear interaction of waves and particles, Phys. Lett. 47A (1974), 165-166.

[13] H. HASHIMOTO, Exact solution of a certain semi-linear system of partial differential equations related to a migrating predation problem, Proc. Japan Acad. 50 (1974), 623-627.

[14] E. HOPF, The partial differential equation $u_t + uu_x = \mu u_{xx}$, Comm. Pure. Appl. Math. 3 (1950), 201-230.

[15] U. d'ANCONA, The Struggle for Existence, E.J. Brill, Leiden, Netherlands, 1954.

[16] K. DAIKOKU AND Y. MIZUSHIMA, New instability concept in avalanche diode oscillation, Japan. J. Appl. Phys. 13 (1974), 989-994.

[17] K. DAIKOKU, Y. MIZUSHIMA, AND T. TAMAMA, Computer experiments on new lattice solitons propagating in Volterra's system, Japan. J. Appl. Phys. 14 (1975), 367-376.

[18] R. HIROTA AND K. SUZUKI, Theoretical and experimental studies of lattice solitons in nonlinear lumped networks, Proc. IEEE 61 (1973), 1483-1491.

[19] M. TODA, Wave propagation in anharmonic lattices, J. Phys. Soc. Japan 23 (1967), 501-506.

[20] V.E. ZAKHAROV AND S.V. MANAKOV, Resonant interaction of wave packets in nonlinear media, Soviet Physics JETP Lett. 18 (1973), 243-245.

[21] K. NOZAKI AND T. TANIUTI, Propagation of solitary pulses in interactions of plasma waves, J. Phys. Soc. Japan 34 (1973), 796-800.

[22] V.S. DRYUMA, Analytic solution of the two-dimensional Korteweg-deVries (KdV) equation, Soviet Physics JETP Lett. 19 (1974), 387-388.

[23] J. SATSUMA, private communication.

[24] R. HIROTA, Exact solution of the Korteweg-deVries equation for multiple collisions of solitons, Phys. Rev. Lett. 27 (1971), 1192-1194.

[25] T.A. FULTON, R.C. DYNES, AND P.W. ANDERSON, The flux shuttle - A Josephson junction shift register employing single flux quanta, Proc. IEEE 61 (1973), 28-35.

[26] R. HIROTA, Exact three-soliton solution of the two-dimensional sine-Gordon equation, J. Phys. Soc. Japan 35 (1973), 1566.

BÄCKLUND TRANSFORMATIONS AT THE TURN OF THE CENTURY

George L. Lamb, Jr.

Department of Mathematics
and
Optical Sciences Center
The University of Arizona
Tucson, Arizona 85721

I. INTRODUCTION

Research on pseudospherical surfaces (sometimes referred to as surfaces of constant negative curvature*; perhaps the simplest example is a bugle-like surface) led A. V. Bäcklund to discover, in about 1875, the transformation theory that now bears his name [1]. Recent successful applications of this transformation theory to nonlinear evolution equations have led to a rekindling of interest in this topic. The following descriptive summary is intended as an introduction to some of its more archaic aspects. A bibliography of papers not referred to in the text has also been appended.

II. PSEUDOSPHERICAL SURFACES

When referred to suitable coordinates on the surface, say u and v, the line element of a surface of constant negative curvature may be written [2]

$$ds^2 = \alpha^2(du^2 + 2\cos\omega\, du dv + dv^2) \tag{1}$$

where $-1/\alpha^2$ is the constant total curvature of the surface and ω, the angle between the asymptotic lines [2], satisfies

$$\frac{\partial^2\omega}{\partial u \partial v} = \sin\omega. \tag{2}$$

To each solution of this equation there corresponds a surface of constant negative curvature. Since (2) is invariant under the one parameter group of transformations $u' = mu$, $v' = v/m$, for each solution $\omega(u,v)$, one may obtain

* A surface of constant curvature is a surface with the same total curvature K (product of principal curvatures) at every point. Then a pseudospherical surface is one with $K < 0$.

solutions $\omega(mu, v/m)$. Such solutions are referred to as being obtained by a transformation of Lie. It was the quest for additional techniques for generating such surfaces that led Bäcklund to the transformation theory that has proven to be so useful. It was found by Bäcklund that a new solution (i.e., surface) ω_1 could be obtained from a given solution ω_0 by means of the relations

$$\frac{\partial}{\partial u}\left(\frac{\omega_1 - \omega_0}{2}\right) = a \sin\left(\frac{\omega_1 + \omega_0}{2}\right) , \tag{3a}$$

$$\frac{\partial}{\partial v}\left(\frac{\omega_1 + \omega_0}{2}\right) = \frac{1}{a} \sin\left(\frac{\omega_1 - \omega_0}{2}\right) , \tag{3b}$$

where a is an arbitrary constant. These equations would be of little practical use if a first solution ω_0 could not be obtained. Fortunately, the solution $\omega_0 = 0$ is evident from inspection and provides a basis for constructing further solutions.

III. THEOREM OF PERMUTABILITY

Although a sequence of solutions can be obtained by using ω_1 as a subsequent choice for ω_0 and then generating a new solution ω_2, etc., a procedure for obtaining new solutions without the use of quadrature is also available. This is known as the theorem of permutability and was discovered by Bianchi [3]. Since (3) represent a transformation from a solution ω_0 to a solution ω_1 with a constant a, they may be represented schematically as shown in Figure 1.

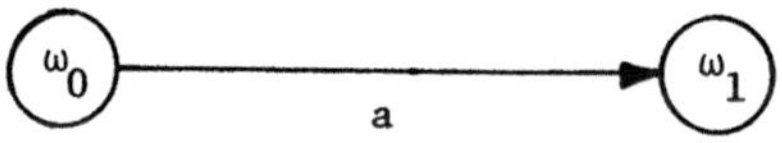

Figure 1. Schematic diagram of Bäcklund transformation given in (3), $\omega_0 \to \omega_1$.

Bianchi showed that four such solutions could be interrelated without the use of quadrature. According to (1) and (3), a surface S characterized by ω_0 can be transformed into a new surface S_1 (of the same curvature) by means of ω_1 and a_1. The theorem of permutability is (cf. [2], p. 286):

If S_1 and S_2 are transforms of S by means of the respective pairs of functions (ω_1, a_1) and (ω_2, a_2), a function ω_3 can be found without quadratures which is such that by means of the pairs (ω_3, a_2) and (ω_3, a_1), the surfaces S_1 and S_2 respectively are transformable into a surface S'.

Only the function ω_3 is of interest here, not the surfaces themselves. Hence the content of the theorem can be indicated schematically as shown in Figure 2.

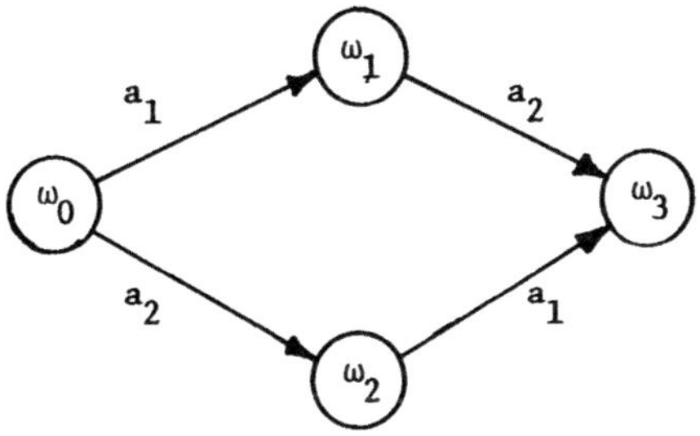

Figure 2. Schematic form of transformations occurring in the theorem of permutability.

The four u-derivative equations associated with Figure 2 are

(4a) $$\frac{1}{2}\frac{\partial}{\partial u}(\omega_1 - \omega_0) = a_1 \sin[\tfrac{1}{2}(\omega_1 + \omega_0)],$$

(4b) $$\frac{1}{2}\frac{\partial}{\partial u}(\omega_2 - \omega_0) = a_2 \sin[\tfrac{1}{2}(\omega_2 + \omega_0)],$$

(4c) $$\frac{1}{2}\frac{\partial}{\partial u}(\omega_3 - \omega_1) = a_2 \sin[\tfrac{1}{2}(\omega_3 + \omega_1)],$$

(4d) $$\frac{1}{2}\frac{\partial}{\partial u}(\omega_3 - \omega_2) = a_1 \sin[\tfrac{1}{2}(\omega_3 + \omega_2)],$$

where ω_3 is the solution referred to above. These four equations can be manipulated algebraically to obtain a form that is completely independent of the u-derivatives. The result may be expressed in the form

(5) $$\tan\left(\frac{\omega_3 - \omega_0}{4}\right) = \frac{a_1 + a_2}{a_1 - a_2}\tan\left(\frac{\omega_1 - \omega_2}{4}\right).$$

Bianchi has shown that if the functions ω_3 in (4c) and (4d) are replaced by, say, ω_4 and ω_5, respectively, then, with ω_1 and ω_2 given by (4a,b), differentiation of (5) implies the validity of (4c,d) with the same ω_3 in both equations.

The theorem of permutability provides an extremely efficient means of generating solutions of (2).

IV. CLAIRIN'S METHOD

Further researches of Bäcklund [4] - [6] led to more general applications of the theory to problems in differential geometry. However, a form of the theory that appears to have been introduced by J. Clairin [7], [8] seems to be somewhat more direct (albeit tedious) in leading to such transformations (when they exist) for arbitrarily chosen equations. To illustrate Clairin's method, let us consider two dependent variables $z(x,y)$ and $z'(x,y)$ and their first derivatives to be interrelated by the pair of equations

(6a) $$p = f(z,z',p',q'),$$

(6b) $$q = \phi(z,z',p',q'),$$

where $p = \partial z/\partial x$, $q = \partial z/\partial y$ and similarly for the primed variables. The integrability condition requires

(7) $$\Omega \equiv \frac{\partial p}{\partial y} - \frac{\partial q}{\partial x} = 0.$$

A fairly complete discussion of (6) when the term z' is absent has been given by Forsyth [9]. In particular he shows quite clearly the conditions under which (6) will reduce to a contact transformation when z' is absent. To see this, note that (7) with z' absent may be written

(8) $$\Omega = f_z\phi - \phi_z f + (f_{p'} - \phi_{q'})s' + f_{q'}t' - \phi_{p'}r' = 0$$

where subscripts denote partial derivatives. We employ the notation

$$r = \partial^2 z/\partial x^2, \quad s = \partial^2 z/\partial x\partial y, \quad t = \partial^2 z/\partial y^2,$$

and similarly for z'. Now, (8) can be satisfied in either of two completely different ways. If r', s', and t' do not appear in (8), i.e., if

(9) $$f_{p'} - \phi_{q'} = f_{q'} = \phi_{p'} = 0$$

then, (6) represent a contact transformation. In that case, as shown in detail by Forsyth, (8) may be transformed to

(10) $$(dz' - p'dx - q'dy) = \mu(dz - pdx - qdy)$$

the standard form for a contact transformation. On the other hand, (8) may also be satisfied when any or all of the relations in (9) are not satisfied. Equation (8) then takes on the form of a second-order partial differential equation of Monge-Ampere form and may be satisfied because z' happens to satisfy this partial differential equation. In this latter case (6) are referred to as a Bäcklund transformation. That not all equations of Monge-Ampere form may play such a role in defining a Bäcklund transformation has been shown by Forsyth (one must be able to satisfy four equations with three unknown functions).

V. A SIMPLE EXAMPLE

As an example of the construction of a Bäcklund transformation using Clairin's method, we consider functions z and z' which satisfy

(11a) $$s' = e^{z'} \quad \text{(Liouville's equation [10])},$$

(11b) $$s = 0 \quad \text{(wave equation)}.$$

We now derive a Bäcklund transformation relating solutions of these two equations.

For the sake of simplicity we shall assume that (6) are of the form

(12a) $$p = p' + m(z,z'),$$

(12b) $$q = -q' + \mu(z,z').$$

If the general form given in (6) were employed as a starting point for the following calculation, a much lengthier analysis would result. Equations (12) anticipate the final result and simplify the analysis. The integrability condition for (12) now imposes the relation

(13) $$\Omega = m_z(-q' + \mu) + m_{z'}q' - \mu_z(p' + m) - \mu_{z'}p' + 2e^{z'} = 0$$

where (11a) has been employed. Also,

(14a) $$\Omega_{p'} = -(\mu_{z'} + \mu_z) = 0,$$

(14b) $$\Omega_{q'} = m_{z'} - m_z = 0.$$

Hence, (13) reduces to

(15a) $$\Omega = \mu m_z - m\mu_z + 2e^{z'} = 0,$$

and hence

(15b) $$\Omega_z = \mu m_{zz} - m\mu_{zz} = 0.$$

By (14), m and μ must be of the form

(16a) $$\mu(z,z') = \mu(z - z'),$$

(16b) $$m(z,z') = m(z + z').$$

If more general expressions for the right sides in (12), were assumed, higher derivatives of Ω would be required before their form could be determined.

Introducing new dependent variables

(17a) $$\zeta = z - z',$$

(17b) $$\eta = z + z',$$

(15b) takes the form

(18) $$\mu(\zeta)\frac{d^2m}{d\eta^2} - m(\eta)\frac{d^2\mu}{d\zeta^2} = 0.$$

Upon separating variables and integrating

(19a) $$m(\eta) = \alpha e^{\kappa\eta} + \beta e^{-\kappa\eta} ,$$

(19b) $$\mu(\zeta) = \gamma e^{\kappa\zeta} + \delta e^{-\kappa\zeta} ,$$

where $\kappa^2 > 0$ is the separation constant (the sign chosen from hindsight - see below) and α, β, γ, and δ are integration constants. Equations (12) now read

(20a) $$p = p' + \alpha e^{\kappa\eta} + \beta e^{-\kappa\eta},$$

(20b) $$q = -q' + \gamma e^{\kappa\zeta} + \delta e^{-\kappa\zeta},$$

and thus far, only (11a) has been employed. Equation (11b) must also be satisfied. The y-derivative of (20a) and use of (20b) lead to

(21) $$s = s' + \kappa(\alpha\gamma e^{2\kappa z} + \alpha\delta e^{2\kappa z'} - \beta\gamma e^{-2\kappa z'} - \beta\delta e^{-2\kappa z}).$$

Equations (11) are satisfied by $\kappa = -\frac{1}{2}$, $\alpha = \delta = 0$, $\gamma = 2/\beta$. Hence the final form of the Bäcklund transformation is,

(22a) $$p = p' + \beta e^{\frac{z+z'}{2}} ,$$

(22b) $$q = -q' - \frac{2}{\beta} e^{-\frac{z-z'}{2}}$$

(An alternative choice would be $\kappa = \frac{1}{2}$, $\alpha = -2/\delta$, $\beta = \gamma = 0$, but the same one parameter Bäcklund transformation would result.)

Equations (22) may be used to obtain the general solution of Liouville's equation. Introducing the general solution of $s = 0$ in the d'Alembert form $z = \theta(x) + \chi(y)$ where θ and χ are arbitrary functions of the indicated variables into (22), the resulting first-order equations are readily integrated. The general solution of Liouville's equation is then found to be [10]

(23) $$e^{z'} = 2 \frac{\theta'(x)\chi'(y)}{[\theta(x) + \chi(y)]^2}$$

where θ' and χ' refer to derivatives of these functions with respect to their arguments.

Since (3) merely transform (2) into itself rather than to an equation for which the general solution is known, the technique only provides particular solutions of (2).

EPILOGUE

Bäcklund transformations ceased to be an active area of research after World War I. Undoubtedly this was due, at least in part, to the demise of many of those active in the field (e.g. Clairin). Later papers by Goursat [11] and Bäcklund [6] appeared and a bibliography from Goursat's paper [11] has been appended here. A truly exhaustive cataloguing of the solutions of (2) that are in the form $\omega = 4\tan^{-1}[F(au + v/a)/G(au - v/a)]$ has been given by Steuerwald [12]. The topic of infinitesimal Bäcklund transformations was considered by Loewner [13]. A paper on transformation theory by Bateman [14] makes brief mention of Bäcklund transformations and includes some references not listed in the bibliography of the present paper. The paper of modern times that seems to have served to rescue the subject from antiquity and paved the way for application of the method in nonlinear pulse propagation is that of Seeger, Donth, and Kochendörfer [15].

The past few years have seen a burgeoning of papers on the subject of Bäcklund transformations. References to them will be found in the other papers of these proceedings.

REFERENCES

[1] A.V. BÄCKLUND, Einiges über Curven und Flächentransformationen, Lund Universitëts Arsskrift 10 (1875), Om Ytor med konstant negativ krökning, Ibid. 19 (1883).

[2] L.P. EISENHART, A Treatise on the Differential Geometry of Curves and Surfaces, Dover Publications, New York, N.Y., 1960, 190.

[3] L. BIANCHI, Lezioni di Geometria Differenziale, Vol. II, Pisa, 1902, 418.

[4] A.V. BÄCKLUND, Math. Ann. 9 (1876), 297-320.

[5] ____________, Zur Theorie der Flächentransformationen, Math. Ann. 19 (1882), 387-422.

[6] A.V. BÄCKLUND, Zur Transformationstheorie partieller differentialgleichungen zweiter Ordnung, Lund Universitëts Arsskrift 31 (1920), 1-33.

[7] J. CLAIRIN, Sur les transformations de Bäcklund, Ann. Sci. École Norm. Sup. 3e Ser. Suppl. 19 (1902), S-1 - S-63.

[8] _________, Sur quelques équations aux dérivées partielles du second ordre, Ann. Fac. Sci. Univ. Toulouse 2e Ser. 5 (1903), 437-458.

[9] A.R. FORSYTH, Theory of Differential Equations, Vol. VI, Chap. 21, Dover Publications, New York, N.Y., 1959.

[10] J. LIOUVILLE, Sur l'equation aux differences partielles $\frac{d^2 \log \lambda}{du\,dv} \pm \frac{\lambda}{2a^2} = 0$, J. Math. Pures Appl. 18 (1853), 71-72.

[11] E. GOURSAT, Le Problem de Bäcklund, Mémor. Sci. Math. Fasc. 6, Gauthier-Villars, Paris, 1925.

[12] R. STEUERWALD, Über Enneper'sche Flächen und Bäcklund'sche Transformation, Abh. Bayerische Akad. Wiss. (Meunchen) 40 (1936), 1-105.

[13] C. LOEWNER, Generation of solutions of systems of partial differential equations by composition of infinitesimal Baecklund transformations, J. Analyse Math. 2 (1952), 219-242.

[14] H. BATEMAN, The transformation of partial differential equations, Quart. Appl. Math. 1 (1944), 281-296.

[15] A. SEEGER, H. DONTH, AND A. KOCHENDORFER, Theorie der Versetzungen in eindimensionalen Atomreihen, Z. Physik 134 (1953), 173-193.

ADDITIONAL REFERENCES FROM [11]

A.V. BÄCKLUND, Zur Theorie der partiellen Differentialgleichungen erster ordung, Mathematische Annalen 17 (1880), 285.

____________, Ueber eine Transformation von Luigi Bianchi, Annali di Matematica, 3e Ser. 23 (1914), 107.

____________, Ein Satz von Weingarten über auf einander abwickelbare Flächen, Lund Universitëts Arsskrift 29 (1918).

L. BIANCHI, Richerche sulle superficie a curvatura costante e sulle elicoidi, Annali della R. Scuola normale superiore di Pisa 2 (1879), 285.

_________, Ueber die Flächen mit constanter negativer Krummung, Mathematische Annalen 16 (1880), 577.

E. CARTAN, Sur l'integration des systèmes d'equations aux differentielles totales, Annales de l'École Normale supérieure, 3e Ser. 18 (1901), 241-311.

________, Les systèmes de Pfaff et les équations aux dérivées partielles du second ordre, Ibid., 3e Ser. 27 (1910), 109-192.

G. CERF, Sur les transformations des équations aux dérivées partielles d'ordre quelconque à deux variables indépendantes, Journal de Mathématiques, 8e Ser. 1 (1918), 309.

J. CLAIRIN, Sur les transformations d'une classe d'équations aux dérivées partielles du second ordre, Annales scientifiques de l'École Normale supérieure, 3e Ser. 27 (1910), 451-489.

__________, Sur quelques points de la théorie des transformations de Bäcklund, Ibid., 3e Ser. 30 (1913), 173-197.

__________, Sur une classe de transformations des équations aux dérivées partielles du second ordre, Bulletin de la Société mathématique 30 (1902), 100-105.

__________, Sur certaines transformations des équations linéaires aux dérivées partielles du second ordre, Ibid. 33 (1905), 90-97.

__________, Sur la transformation d'Imschenetsky, Ibid. 41 (1913), 206-228.

__________, Sur une transformation de Bäcklund, Bulletin des Sciences mathématiques, 2e Ser. 24 (1900), 284.

__________, Sur les transformations de quelques équations linéaires aux dérivées partielles du second ordre, Mémoire posthume, Annales de l'Ecole Normale, 3e Ser. 37 (1920), 95.

E. COSSERAT, Sur la déformation de certains paraboloïdes et sur le théorème de M. Weingarten, Comptes rendus 124 (1897), 741.

G. DARBOUX, Lecons sur la théorie générale des surfaces, 2, Livre IV, Chap. II, V, VI, VII, VIII, IX; 3, Livre VII, Chap. XII et XIII.

__________, Sur la déformation des surfaces du second degré et sur les transformations des surfaces à courbure totale constante, Comptes rendus 128 (1899), 760, 854, 953, 1018.

M. DUPORT, Mémoire sur les équations différentielles, Journal de Mathématiques pures et appliquées, 5e Ser. 3 (1897), 17.

E. GAU, Sur les transformations les plus générales des équations aux dérivées partielles du second ordre, Comptes rendus 156 (1913), 116.

E. GOURSAT, Lecons sur le problème de Pfaff, Paris, 1923.

__________, Lecons sur l'intégration des équations aux dérivées partielles du second ordre, 1, Chap. II et IV; 2, Chap. IX.

__________, Sur une équation aux dérivées partielles, Bulletin de la Société mathématique 25 (1897), 36.

__________, Sur une transformation de l'équation $s^2 = 4\lambda pq$, Ibid. 28 (1900), 1.

__________, Sur quelques transformations des équations aux dérivées partielles du second ordre, Annales de la Faculté de Toulouse, 2e Ser. 4 (1902), 299-340.

__________, Sur le problème de Bäcklund et les systèmes de deux équations de Pfaff, Ibid. 3e Ser. 10 (1918).

__________, Sur quelques équations du second ordre qui admettent une transformation de Bäcklund, Bulletin de la Société mathématique 49 (1921), 1-65.

__________, Sur les éléments singuliers d'un système de deux équations de Pfaff, Ibid. 52 (1924), 38.

E. GOURSAT, Sur quelques transformations d'équations aux dérivées partielles, Bulletin des Sciences mathématiques, 2e Ser. 46 (1922), 370.

_________, Sur quelques équations aux dérivées partielles de la théorie de la déformation des surfaces, Comptes rendus 180 (1925), 1303.

_________, Sur quelques transformations des équations aux dérivées partielles du second ordre, Comptes rendus 170 (1920), 1217.

_________, Sur une classe d'équations aux dérivées partielles du second ordre et sur la théorie des intégrales intermédiaires, Acta mathematica 19 (1895), 285.

V.G. IMSCHENETSKY (traduit par Houel), Étude sur les méthodes d'intégration des équations aux dérivées partielles du second ordre à deux variables indépendantes, Archives de Grunert 54, 257.

L. LEVY, Sur quelques équations linéaires aux dérivées partielles du second ordre, Journal de l'Ecole Polytechnique, 56e cahier (1886).

S. LIE, Ueber Flächen deren Krummungs adien durch eine relation verknüpft sind, Archiv for Mathematik og Naturvidenskab 4 (1879), 510.

R. LIOUVILLE, Sur les équations aux dérivées partielles du second ordre, qui contiennent linéairement les dérivées de l'ordre le plus élevé, Comptes rendus 98 (1884), 216, 569, 723.

MOUTARD, Sur la construction des équations de la forme $\frac{\partial^2 z}{\partial x \partial y} = \lambda(x,y)z$, qui admettent une intégrale générale explicite, Journal de l'Ecole Polytechnique, 45e cahier (1878).

E. PICARD, Sur une généralisation des équations de la théorie des fonctions d'une variable complexe, Comptes rendus 112 (1891), 1399.

_________, Sur certains systèmes d'équations aux dérivées partielles, Ibid. 114 (1892), 805.

G. TEIXEIRA, Sur l'intégration d'une classe d'équations aux dérivées partielles du second ordre, Bulletin de l'Académie de Belgique, 3e Ser. 3 (1882), 486.

THE APPLICATION OF BÄCKLUND TRANSFORMS TO PHYSICAL PROBLEMS*

Alwyn C. Scott†

Department of Electrical and Computer Engineering
University of Wisconsin
Madison, Wisconsin 53706

I. INTRODUCTION

Bäcklund transform techniques have been used to find some solutions for some partial differential equations. Currently it is interesting to learn how they are used and for which equations. The purpose of these notes is to record some results and open questions which may be of interest to participants in the NSF Workshop, and more generally to the scientific community at large. Without lapsing into extended discussions of experimental science, an effort is made to relate the pde's considered to the physical problems from which they arise. Since various forms of the nonlinear Klein-Gordon equation

$$\nabla^2\phi - \phi_{tt} = F(\phi) \tag{1}$$

have been of direct experimental interest to me for the past eight years [1]-[5], it necessarily assumes a central role in the discussion.

II. LINEAR PARTIAL DIFFERENTIAL EQUATIONS

For the large area Josephson junction [5], ∇^2 in (1) is the Laplacian in two-space dimensions; and, since I am interested in localized solutions, it is convenient to write (1) in polar coordinates as

$$r^2\phi_{rr} + r\phi_r + \phi_{\theta\theta} - r^2\phi_{tt} = r^2F(\phi) \ . \tag{2}$$

---

*This work was partially supported by the Office of Naval Research under Contract No. N00014-67-A-0467-0027 at the Applied Mathematics Summer Institute and by the National Science Foundation under Grant No. GK-37552.

†Presently at Mathematics Research Center, University of Wisconsin, Madison, Wisconsin 53706.

Given a solution ϕ, its differential can be written as the Pfaffian form

(3) $$d\phi = Pdr + Qd\theta + Rdt$$

where $\phi_r = P$, $\phi_\theta = Q$, and $\phi_t = R$. It is interesting to consider how P, Q, and R might be chosen to insure that ϕ is a solution. For the linear assumption

(4) $$F(\phi) = \phi ,$$

let us take P, Q, and R in the form

(5a,b,c) $$P = G(r)\phi , \; Q = b\phi , \; R = c\phi$$

where b and c are constants. Then (2) requires that G satisfy the first-order Riccati equation [6]

(6) $$G' + G^2 + \frac{1}{r} G = 1 + c^2 - b^2/r^2 .$$

With the "standard linearizing substitution" $G = \Phi_r/\Phi$, the corresponding linear equation is Bessel's equation

(7) $$\Phi_{rr} + \frac{1}{r}\Phi_r + (\frac{b^2}{r^2} - 1 - c^2)\Phi = 0 .$$

Since we are considering (2) to be linear, we can interpret

(8) $$\phi = \phi_n - \phi_{n-1}$$

where ϕ_{n-1} is an "old" solution and ϕ_n is a "new" solution being sought. Then (5) become the Bäcklund transform

(9a) $$\phi_{n,r} = \phi_{n-1,r} + G(r)(\phi_n - \phi_{n-1}) ,$$

(9b) $$\phi_{n,\theta} = \phi_{n-1,\theta} + b(\phi_n - \phi_{n-1}) ,$$

(9c) $$\phi_{n,t} = \phi_{n-1,t} + c(\phi_n - \phi_{n-1}) .$$

Each iterated integration of (9) (i.e., each "turn of the Bäcklund crank") introduces an additional radial eigenfunction into the solution for which the appropriate angle, time, and radial dependences are assured by selection of the constants b and c and the function G(r).

A corresponding discussion was previously carried through in detail for the one-dimensional, linear Klein-Gordon equation [7]

$$\phi_{xx} - \phi_{tt} = k + \phi \,, \tag{10}$$

but it seems that the procedure works for any separable, linear pde. Spatial dependence, as in the linear Schrödinger equation

$$\frac{1}{2}(\phi_{xx} - x^2\phi) = i\phi_t \,, \tag{11}$$

merely requires a corresponding explicit spatial dependence of the Bäcklund transform just as in (5a). Higher than second-order derivatives lead to "generalized Riccati equations" of higher order [6] for which the order of the associated linear problem is correspondingly increased. Dependence upon more than one independent variable requires more than one Riccati equation. For example, in a spherical coordinate system the linear Klein-Gordon equation generates two Riccati equations one of which governs the radial dependence and the other the latitude dependence of the spherical harmonics.

But, as Gerber [8] emphasizes, one should recognize the importance of Bäcklund transforms exhibiting explicit dependence on the independent variables which is not shared by the original pde. For example, the "raising" operator for eigenfunctions of (11) can be readily expressed as the Bäcklund transform

$$\phi_{n,x} = \left(\frac{\partial^2}{\partial x^2} - x\frac{\partial}{\partial x} - 1\right)\phi_{n-1}e^{it} \,, \tag{12a}$$

$$\phi_{n,t} = i(n + \tfrac{1}{2})\phi_n \,. \tag{12b}$$

If the "old" solution, ϕ_{n-1}, is an eigenfunction of (11), the "new" solution, ϕ_n, will be the next higher eigenfunction. This is similar to the Bäcklund transform developed by Rogers [9] for generating a hierarchy of solutions to the axially symmetric, incompressible flow problem in hydrodynamics. The raising and lowering operators of quantum mechanics are also closely related to the "generative operators" developed by Moseley [10] for obtaining nonseparable solutions to the Helmholtz equation

$$\nabla^2\phi + \phi = 0 . \tag{13}$$

As Schoonaert and Luypaert [11] have recently demonstrated, certain of these nonseparable solutions are useful in the design of microwave structures with improved capability to store energy and transport power.

III. NONLINEAR DIFFUSION EQUATIONS

1. The Nerve Problem

A most important problem in electrophysiology is the nonlinear diffusion of electrical potential along an active nerve fiber (axon) according to the equation

$$\phi_{xx} - \phi_t = I_i \tag{14}$$

where ϕ is electrical potential (voltage) across the surface of a cylindrical membrane, I_i is ion current through the membrane, and x is distance along the cylinder axis [12]. In general, I_i is a rather complicated nonlinear function involving several auxiliary variables; but during the first rapid rise (leading edge) of a nerve pulse, it can be approximated as a function only of membrane voltage, $F(\phi)$, so (14) becomes

$$\phi_{xx} - \phi_t = F(\phi) . \tag{15}$$

The nonlinear function $F(\phi)$ goes through zero with positive slope at two values of ϕ (say ϕ_1 and ϕ_3) and with negative slope at ϕ_2 where $\phi_1 < \phi_2 < \phi_3$. Thus, in general, $F(\phi)$ has "cubic" behavior for $\phi_1 \leq \phi \leq \phi_3$.

It is of interest to determine the shape and the velocity of the leading edge of a nerve pulse from analysis of (15). With the assumption that ϕ is a traveling wave of velocity u

$$\phi(x,t) = \Phi(x - ut) \ , \tag{16}$$

Bäcklund transform equations which will generate traveling wave solutions from the vacuum $(\phi = 0)$ solution are

$$\Phi_x = P_0(\Phi) \tag{17a}$$

$$\Phi_t = -uP_0(\Phi) \ . \tag{17b}$$

From (15), P_0 must satisfy

$$P_0(P_0' + u) = F(\Phi) \ . \tag{18}$$

To appreciate the physical character of these traveling waves, consider the ordinary candle. The fully developed flame is a stable wave while the threshold condition for ignition is determined by an unstable wave of much smaller velocity.

Assuming $u = 0$ in (18) we have

$$P_0 = \pm\left[2\int_{\phi_1}^{\Phi} F(\alpha)\,d\alpha\right]^{\frac{1}{2}} \tag{19}$$

which integrates to the unstable threshold pulse [12]. For $u \neq 0$ we must solve

$$P_0' = \frac{F}{P_0} - u \tag{20}$$

in order to find the faster wave. Now if P_0 and F are polynomials of degree m and n, respectively, then (20) is satisfied only if

$$n = 2m - 1 \ . \tag{21}$$

Choosing $m = 1$ implies $n = 1$, the "piecewise linear" approximation for $F(\phi)$ shown in Figure 1 which has proved useful in making first-order estimates of nerve pulse velocity [13]. Choosing $m = 2$ implies $n = 3$, a true cubic form

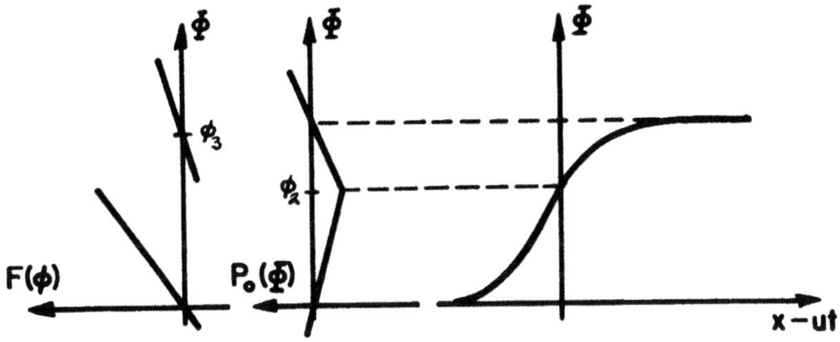

Figure 1. Piecewise linear approximation for F(Φ).

for F(Φ) which leads to the velocity formula that Nagumo ascribes to Huxley [14],[15,pp. 186-190]. Choosing larger values for m or other functional forms for P_0 and F leads to other velocity formulas. To my knowledge, these extensions are largely unexplored.

2. The Burgers Equation

The Burgers equation [16;Chap. 4]

$$\phi_{xx} + 2\phi\phi_x = \phi_t \tag{22}$$

is a nonlinear diffusion equation with a conservation law. Such a conservation law always implies an associated linear problem. To see this assume

$$[f(\phi)]_x = [g(\phi)]_t \tag{23}$$

and construct a Pfaffian form corresponding to $\log\psi$ by

$$d(\log\psi) = (g + \lambda)dx + fdt \tag{24}$$

where λ is independent of time. Then the conservation law (23) assures integrability for $\log\psi$ and we can write

$$\psi_x = [g(\phi) + \lambda]\psi \; , \tag{25a}$$

$$\psi_t = f(\phi)\psi \; . \tag{25b}$$

For (22) and assuming $\lambda = 0$, (25a) becomes

(26a) $$\psi_x = \phi\psi ,$$

the Cole-Hopf transformation under which (25b) reduces to

(26b) $$\psi_t = \psi_{xx} .$$

The Bäcklund transform which corresponds to the Burgers equation (22) can be found from the Bäcklund transform for (26b)

(27a) $$\psi_{n,x} = \psi_{n-1,x} + k(\psi_n - \psi_{n-1}) ,$$

(27b) $$\psi_{n,t} = \psi_{n-1,t} + k^2(\psi_n - \psi_{n-1}) ,$$

and then using (26a) to get back to ϕ. Thus

(28a) $$\phi_{n,x} = -\phi_n^2 + k\phi_n + (\phi_{n-1,x} + \phi_{n-1}^2 - k\phi_{n-1})H$$

(28b) $$\phi_{n,t} = -k\phi_n^2 + k^2\phi_n + [(\phi_n + \phi_{n-1})(\phi_{n-1,x} + \phi_{n-1}^2)$$
$$-k\phi_{n-1}(k + \phi_n) + \phi_{n-1,t}]H$$

where

(29) $$H \equiv \exp\left[\int^x (\phi_{n-1} - \phi_n)\partial x' + \int^t (\phi_{n-1,x} + \phi_{n-1,x}^2 - \phi_{n,x} - \phi_n^2)\partial t'\right] .$$

This is clearly an example in which it is more useful to know the linearizing Cole-Hopf transformation than the Bäcklund transform.

IV. NONLINEAR KLEIN-GORDON EQUATIONS

1. One-Dimensional Josephson (Superconducting) Transmission Line

Currently we are using previously developed integrated circuit techniques [17] to make Josephson junction transmission lines as indicated in Figure 2. The sketch in Figure 3 emphasizes that the primary dynamic effect is *propagation of magnetic flux quanta* in the longitudinal (x) direction. This

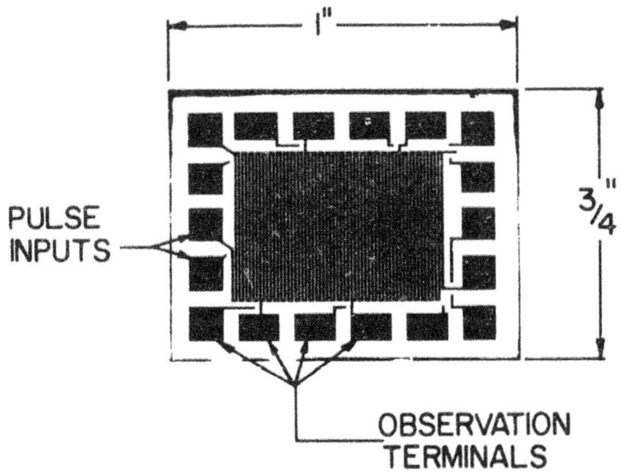

Figure 2. Seventy cm. superconductive transmission line on a 1"×3/4" glass substrate. A 0.012 cm. wide lead strip overlays a 0.048 cm. wide niobium strip with an intervening barrier layer of niobium oxide.

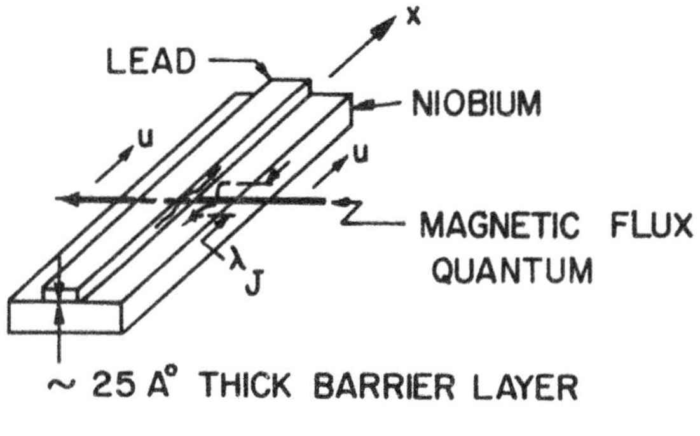

Figure 3. Detail of the superconductive transmission line.

magnetic flux penetrates the very thin (~25 Å) barrier layer of niobium oxide which separates the two superconducting metals (niobium and lead) and is directed perpendicular to x.

The simplest representation of magnetic flux dynamics is given by the one-dimensional sine-Gordon equation [4]

(30) $$\phi_{xx} - \phi_{tt} = \sin\phi \ .$$

In writing (30), velocity has been normalized to a value u_J which is equal to about 1/20 of the free space velocity of light, and distance has been normalized to a Josephson length

$$\lambda_J \approx 0.1 \quad \text{to} \quad 0.01 \ \text{mm} \ .$$

The dependent variable ϕ is related to the magnetic flux Φ by

(31) $$\phi = 2\pi\Phi/\Phi_0$$

where $\Phi_0 = h/2e$ is the flux quantum. In laboratory units

(32) $$\Phi_0 = 2.06 \times 10^{-15} \quad \text{volt-seconds.}$$

A voltage pulse of amplitude V and duration τ would carry $V\tau$ volt-seconds of magnetic flux or $V\tau/\Phi_0$ flux quanta. If pulse measurements were made with an oscilloscope having a voltage sensitivity of 10^{-4} volts and a time resolution of 10^{-9} seconds, the smallest observable pulse would contain about fifty flux quanta. Single flux quanta have been detected, however, in the "Josephson junction shift register" described by Fulton, Dynes, and Anderson [18].

The ϕ_{xx} term in (30) is related to magnetic energy storage in the flux quanta, the ϕ_{tt} term to electric energy in the niobium oxide barrier layer, and the $\sin\phi$ term to a quantum contribution to the free energy.

Several simplifying approximations have been made in the derivation of (30). Probably the most important is neglect of mechanisms for energy dissipation, but also the $\sin\phi$ is an idealization. Physical considerations only require that the right-hand side be an odd periodic function.

2. One-Dimensional Nonlinear Klein-Gordon Equations

Equation (30) which was rather extensively reviewed a few years ago [5], [19], finds application also to the propagation of Bloch walls between magnetic domains and of crystal dislocations [20]. A closely related equation

(33) $$\phi_{xt} + \phi_{tt} = \sin\phi \ ,$$

which has been carefully considered by Lamb [21] in connection with the <u>self induced transparency</u> (SIT) problem of nonlinear optics, also arises in the description of the traveling wave parametric amplifier [22]. Additional references are recorded in a recent review of soliton theory [23].

Equations (30) and (33) can both be put into the canonical form

(34) $$\phi_{xy} = \sin\phi$$

for which the Bäcklund transform

(35) $$\begin{aligned} \phi_{n,x} &= \phi_{n-1,x} + 2a \sin\left(\frac{\phi_n+\phi_{n-1}}{2}\right) \\ \phi_{n,y} &= -\phi_{n-1,y} + \frac{2}{a} \sin\left(\frac{\phi_n-\phi_{n-1}}{2}\right) \end{aligned}$$

is suggested in an exercise in Forsyth [24] with references to Bianchi and to Darboux.

Each iteration of (35) introduces an additional soliton into the solution and, for the Josephson superconducting transmission line idealized as in (30), this soliton represents a quantum of magnetic flux. In isolation it appears as the gudermannian

(36) $$\phi = 4 \tan^{-1}\left[\exp\left(\frac{x - ut + \delta}{\sqrt{1 - u^2}}\right)\right]$$

where the velocity

(37) $$u = \frac{1 - a^2}{1 + a^2}$$

is determined by the choice of the constant a in (35) and δ is a constant of integration. Lamb [21] has shown how (35) may be used to construct "N-soliton" solutions which, for a Josephson superconducting transmission line, correspond to propagating bundles of flux [4].

Bäcklund transforms have been obtained for the more general equation [7]

(38) $$\phi_{xy} = F(\phi)$$

where $F(\phi)$ satisfies the condition

(39) $$F'' \propto F \ .$$

Ablowitz, Kaup, Newell, and Segur [25] and Chu and Scott [26] have established the relation between these Bäcklund transforms and the corresponding linear equations of the inverse scattering transform method. This is accomplished by writing the Bäcklund transforms as Riccati equations for which the associated linear equations can be interpreted as the inverse scattering equations.

3. Active Josephson Transmission Line

An augmentation of (30) to include mechanisms for energy input and loss takes the form [15]

(40) $$\alpha\phi_{xxt} + \phi_{xx} - \ell c\phi_{tt} = \beta\phi_t + \sin\phi - j_B/j_0 \ .$$

Normal electron current flow parallel to and across the insulating barrier are represented respectively by the terms of $\alpha\phi_{xxt}$ and $\beta\phi_t$. The term j_B/j_0 represents a normalized bias current which can provide energy input by applying a Lorentz force (equal to $j_B\Phi_0$) to each flux quantum. The parameter $[\ell c]^{-\frac{1}{2}}$ is about 1/20 the velocity of light as was noted above.

Equation (40) has traveling wave solutions consisting of bundles of flux quanta. Typical solutions for four quanta are indicated in Figure 4 together with a locus of allowed values for normalized bias current density (j_B/j_0). The waveforms plotted are the pulse voltage (v) which, from Faraday's law, is proportional to ϕ_t. The numerical data were obtained on a hybrid (analog-digital) computer [27] by integrating the ordinary differential equation to which (40) reduces under the assumption

(41) $$\phi = \tilde{\phi}(x - ut) \ .$$

Figure 4 indicates that the wave at point C (in the (u,j_B)-plane) has a flux quantum which is becoming detached from the trailing edge. Recently Nakajima et al [28], [29] have studied (40) by a mechanical analog (using air nozzles to provide the energy input!) and also by numerical simulation. Their results indicate that the flux quanta are mutually attractive so the high velocity branch in Figure 4 (points A and B) is stable while the low velocity branch (point C) is unstable.

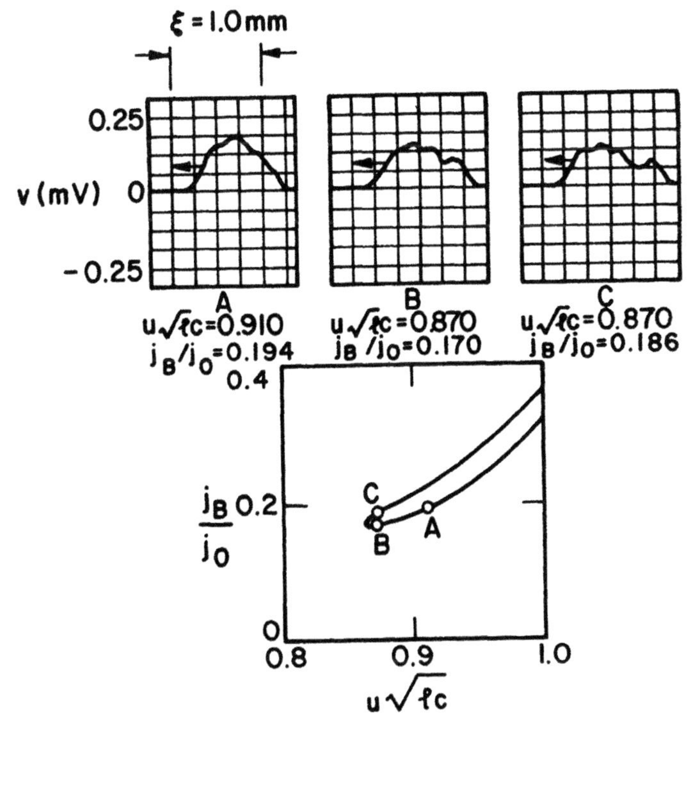

Figure 4. Solitary waves on the active Josephson transmission line.

Equation (40) is closely related to the FitzHugh-Nagumo equation for pulse propagation on a nerve axon [12]. On active lines of this class, waves propagate at the fixed velocity for which power dissipated by the pulse is equal to the energy released per unit time by the nonlinearity. Consider an ordinary candle. Suppose the flame requires P joules/second of power to keep it glowing, and suppose E joules/meter of chemical energy is stored in the wax. Then a <u>power balance condition</u>

$$P = uE \tag{42}$$

fixes the propagation velocity u. Condition (42) is satisfied along the (u,j_B)-locus in Figure 4, and I surmise that $uE > P$ inside since the low velocity branch is unstable.

I feel that (40) is likely to play an important role in clarifying the dependence of the soliton properties of (30) upon structural changes in the pde's. With $\alpha = 0$ but $\beta \neq 0$ and $j_B \neq 0$, only single flux quanta can travel at the fixed velocity satisfying (42) [27]. With $\alpha \neq 0$, $\beta \neq 0$ and $j_B \neq 0$, the flux quanta are attractive [28], [29]. I don't yet know whether this attraction requires $j_B \neq 0$.

4. Two-Dimensional Josephson Junction

The restriction to a single space dimension in Figure 2 and in (30) is only for experimental and analytical convenience. In general it would be interesting to understand the dynamics of a Josephson <u>surface</u> described by the two-dimensional sine-Gordon equation

$$\nabla^2\phi - \phi_{tt} = \sin\phi \tag{43}$$

since such a surface can be readily constructed [17]. On this surface, the voltage (v) across the insulating barrier would be proportional to ϕ_t and a surface current vector $(\bar{i} = i_x\bar{a}_x + i_y\bar{a}_y)$ would be proportional to $\text{grad}_2\phi$ (where grad_2 is a two-dimensional gradient operator) [2].

Equation (43) is invariant under a Lorentz transformation, i.e., for constant velocity u in the x direction the transformation

(44a) $$x \to x' \equiv \frac{x - ut}{\sqrt{1 - u^2}}$$

(44b) $$y \to y' \equiv y$$

(44c) $$t \to t' \equiv \frac{t - ux}{\sqrt{1 - u^2}}$$

(44d) $$\phi \to \phi' \equiv \phi$$

leaves (43) unchanged. Thus it is not necessary to analyze (43) for states of steady translational motion since such motion can always be obtained from a stationary solution by changing the independent variables as in (44).

Time independent solutions of (43) can be obtained by writing it in rectangular coordinates as

(45) $$\phi_{xx} + \phi_{yy} = \sin\phi \ ,$$

and choosing those solutions of (30) for which ϕ remains real when t is replaced by iy. Such static solutions can then be set in uniform motion through the Lorentz transformation (44). Somewhat more general solutions of this class have been presented by Hirota [30]. Some numerical results have been presented by Nakajima, et al [29].

From a more general point of view the problem is to understand the dynamics of flux lines on the Josephson surface. To appreciate the nature of these lines note that (43) can be derived from the energy density

(46) $$H = \tfrac{1}{2}[|\mathrm{grad}_2\phi|^2 + \phi_t^2] + (1 - \cos\phi)$$

with maxima along lines where $\phi = (2n + 1)\pi$ and going to zero exponentially within a distance $O(\lambda_J)$ on either side. Thus a flux line is like a "rubber band" in that it absorbs energy when it gets longer and gives it up as it gets shorter.

Let us seek solutions which consist of flux lines closed on themselves with radial symmetry. This brings us back to (2) with $F(\phi) = \sin\phi$, $\phi_{\theta\theta} = 0$, and $\phi_{tt} = 0$ or

$$\phi_{rr} + \frac{1}{r}\phi_r = \sin\phi \ . \tag{47}$$

Assuming a "vacuum" Bäcklund transform of the form

$$\phi_r = P_0(\phi, r) \tag{48}$$

$$P_0 \frac{\partial P_0}{\partial \phi} + \frac{\partial P_0}{\partial r} + \frac{1}{r} P_0 = \sin\phi \ . \tag{49}$$

This is not much help because (49) seems more difficult to solve than (47). Numerical solution of (47) [5] leads to a logarithmic singularity at the origin of the form

$$\phi = a \log r \tag{50}$$

$$a = \frac{1}{2\pi}\int \mathrm{div}_2 \mathrm{grad}_2 \phi \ d(\mathrm{area}) = \frac{1}{2\pi}\int \sin\phi \ d(\mathrm{area}) \ . \tag{51}$$

In physical terms the fields fall off exponentially for large r so all the Josephson current which crosses the junction accumulates at $r = 0$. Such a boundary condition can only be satisfied if the junction has a "short circuit" at the origin. But a short circuit cannot move; thus this solution cannot be Lorentz transformed into a state of motion.

5. The "Sawtooth-Gordon" Equation

A comparison of the results in the previous section with those in Section II is instructive. With $F(\phi) = \phi$, any finite set of eigenfunctions can be generated through a finite number of Bäcklund iterations. With $F(\phi) = \sin\phi$ and a cylindrical coordinate system, a vacuum Bäcklund transform for even the most simple solution has not yet been found. This stark contrast has led Dave McLaughlin and me to consider the piecewise linear or "sawtooth" approximation to the sine function indicated in Figure 5.

Such an approximation is useful for a description of the leading edge of a nerve pulse (see Figure 1) and it readily yields a smooth solution to

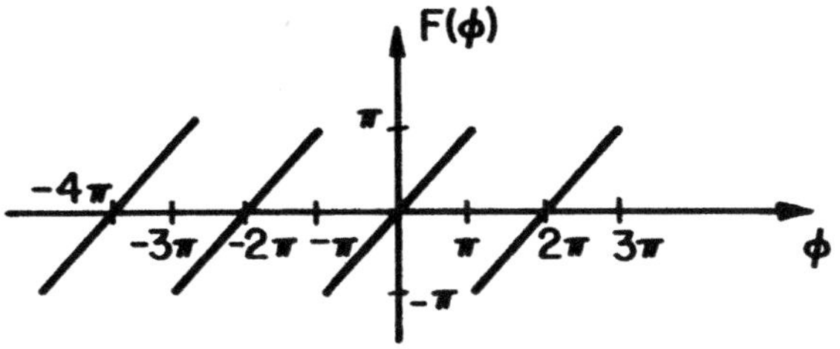

Figure 5. The "sawtooth" approximation to $\sin\phi$.

$\phi_{xx} = f(\phi)$ as

$$\phi = \begin{cases} \pi e^{x} & (x \leq 0) \\ 2\pi - \pi e^{-x} & (x \geq 0) \end{cases} \tag{52}$$

which corresponds to a stationary form of (36). We are planning a systematic study of (1) with $F(\phi)$ as in Figure 5. Some preliminary results are indicated in the next two sections.

6. Wave Motion on Flux Lines

Since flux lines are like rubber bands, we might expect them to support waves. With the "sawtooth" representation of $F(\phi)$, an equation for these waves is readily obtained. Consider Figure 6 where a flux line is assumed to

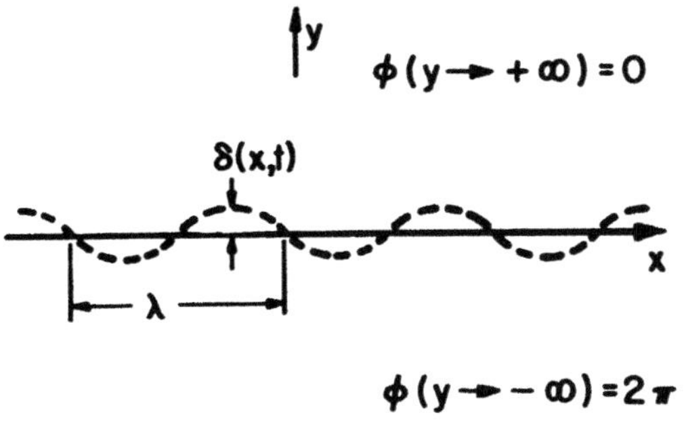

Figure 6. Wave motion on a flux line.

lie among the x-axis and $\delta(x,t)$ is the displacement of the line where $\phi = \pi$. For $\delta = 0$, reference to (52) tells us that a solution is

$$\phi = \begin{cases} 2\pi - \pi e^{y} & (y \leq 0) \\ \pi e^{-y} & (y \geq 0) \end{cases} \tag{53}$$

Now if $\delta \neq 0$ but sufficiently small compared with the wavelength (λ), corresponding solutions are

$$\phi = \begin{cases} 2\pi - \pi e^{y-\delta} & (y \leq \delta) \\ \pi e^{-y+\delta} & (y \geq \delta) . \end{cases} \tag{54}$$

Substitution into $\phi_{xx} + \phi_{yy} - \phi_{tt} = F(\phi)$ yields the wave equation

$$\delta_{xx} - \delta_{tt} + \delta_x^2 - \delta_t^2 = 0 . \tag{55}$$

Equation (55) has the dispersion equation $\omega = \pm k$ so one expects it to be closely related to a linear equation. Since it is Riccati in both δ_x and δ_t, the suggested transformation $\delta = \log G$ linearizes it to $G_{xx} = G_{tt}$.

7. The Rotation Dipole

If $\phi \to 0$ as $r \to \infty$, the "sawtooth" representation implies $F(\phi) = \phi$ in the outer region. One interesting form for the outer solution is then

$$\phi = \frac{k}{r} \cos(\theta - r) \tag{56}$$

which "spins" at $\omega = 1$ (the Josephson frequency in unnormalized units). In a rotating plane of r,θ (where $\theta' = \theta - t$), the lines where $\theta = \pm\pi$ are easily sketched as indicated in Figure 7. These are important lines to consider for inside them (see Figure 5) we must switch to new branches of $F(\phi)$. Since Cauchy conditions are specified on the boundaries, we should expect to have singularities in the shaded regions. But under the transformation

$$r = 1/\rho \tag{57}$$

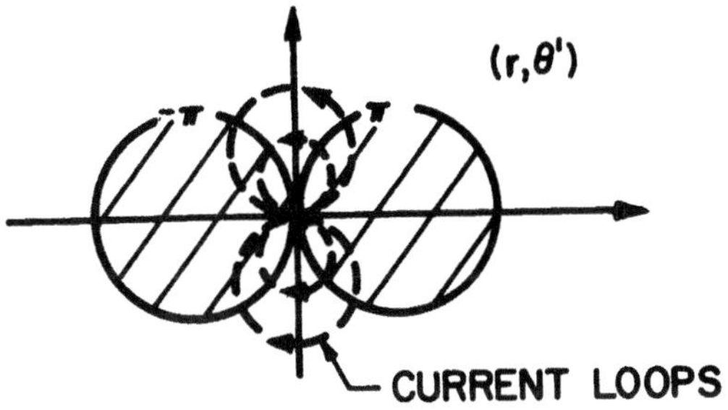

Figure 7. Sketch of a localized solution in the $r - \theta'$ plane.

(1) becomes

(58) $$[(\xi^2+\eta^2)^2 - \eta^2]\phi_{\xi\xi} + [(\xi^2+\eta^2)^2 - \xi^2]\phi_{\eta\eta} + \xi\phi_\xi + \eta\phi_\eta = F(\phi)$$

where

(59a) $$\xi \equiv \rho\cos\theta' \; ,$$

(59b) $$\eta \equiv \rho\sin\theta' \; .$$

Furthermore, (as in Figure 8) the $\phi = \pm\pi$ boundaries transform to the vertical lines $\xi = \pm\,\pi/k$ along which grad$\phi = \phi_\xi = \pm\,k$ and the shaded region is outside.

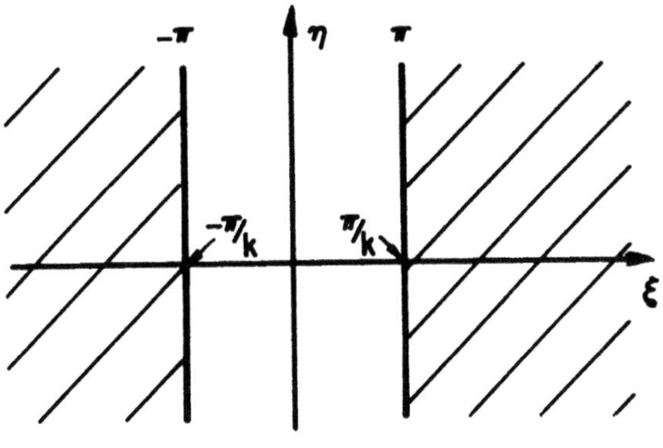

Figure 8. Reciprocal transformation of Fig. 7.

For small k, $|\rho|$ is large in the shaded region and (58) is approximately Laplace's equation $\phi_{\xi\xi} + \phi_{\eta\eta} \approx 0$. The boundary conditions then imply $\phi \approx k\xi$ in the shaded regions so $\phi \to \pm\infty$ as $\xi \to \pm\infty$. Back in Figure 7 this implies a doublet singularity at the origin which is associated with the concentration of current loops. Such a solution may be of experimental interest as a diagnostic probe to monitor the uniformity of large Josephson junction sheets [17] before committing them to integrated circuit processing.

8. Three Space Dimensions

The same games can be played in three space dimensions with additional flexibility. The dynamics of <u>surfaces</u> where $\phi = (2n+1)\pi$ are of interest and such surfaces could be cylindrical, toroidal, spherical, etc., and like "rubber membranes" would support boundary (surface) wave motion analogous to those on flux lines. For references to the application of these solutions to the description of elementary particles, consult the bibliographies in [5] and [19].

V. BOUSSINESQ EQUATIONS

1. General

In 1872 Boussinesq [31] derived the equation

(60) $$\phi_{xx} - \phi_{tt} = -[(\phi^2)_{xx} + \tfrac{1}{4}\phi_{xxxx}]$$

to describe the propagation of shallow water waves. In connection with a study of the Fermi-Pasta-Ulam recurrence phenomenon, Zabusky [32] studied a nonlinear lattice in the continuum approximation and obtained an equation which can be put in the form of (60) [33]. (<u>Note</u>: The lecture notes by Toda [33] are an important recent addition to the soliton literature.) Hirota [34] has obtained N-soliton formulas for (60). A related equation for an electrical transmission line with nonlinear capacitance has recently been studied by Lonngren, Hsuan, and Ames [35]. In the following section I sketch how the Boussinesq equation may be derived for ion sound waves in a plasma.

2. Ion Sound Waves in a Plasma

Starting with

(61) Poisson's equation: $\psi_{xx} = e^{\psi} - n$,

(62) Conservation of ions: $n_t + (nv)_x = 0$,

(63) Newton's second law: $v_t + vv_x = -\psi_x$,

where ψ is electrical potential, n is ion density, v is ion velocity in the x direction, and all physical constants have been set equal to unity, a zero-order solution is evidently

(64) $$n = 1, \quad v = 0, \quad \psi = 0 .$$

Writing $n = 1 + n_1$ and keeping only linear terms in (61)-(63) leads to an equation for the first-order approximation of

(65) $$\psi_{xx} - \psi_{tt} = -\psi_{xxtt} .$$

Assuming the dispersive correction on the right side is small, it can be evaluated by the approximation $\psi_{xx} \approx \psi_{tt}$ so (65) is approximated by

(66) $$\psi_{xx} - \psi_{tt} = -\psi_{xxxx}$$

which differs from the linear part of (60) only by a factor of 1/4 which is easily introduced through a scale change.

For a second-order approximation we keep quadratic terms in (61)-(63) but use the first-order result to evaluate them. Then in the long wave approximation $n_1 = \psi$ and

(67) $$\psi_{xx} - \psi_{tt} = -\psi_{xxtt} - \frac{3}{2} (\psi^2)_{xx} + \frac{1}{2}(\psi^2)_{tt}$$

which can be rearranged to give

(68) $$(\psi + \tfrac{1}{2}\psi^2)_{xx} - (\psi + \tfrac{1}{2}\psi^2)_{tt} = -[(\psi^2)_{xx} + \psi_{xxxx}] .$$

Defining

(69) $$\phi \equiv \psi + \tfrac{1}{2}\psi^2 \ ,$$

ignoring the quadratic correction in the dispersive and nonlinear terms on the right side of (68), and again assuming $\psi_{xx} \approx \psi_{tt}$, we obtain

(70) $$\phi_{xx} - \phi_{tt} = -[(\phi^2)_{xx} + \phi_{xxxx}] \ .$$

Recently ion acoustic solitons have been observed in a cylindrical geometry [36] and numerically studied in a spherical geometry [37], so it is interesting to consider the generalization of (70) to three space dimensions. The basic equations become

(71) $$\nabla^2\psi = e^\psi - n \ ,$$

(72) $$n_t + \mathrm{div}(n\bar{v}) = 0 \ ,$$

(73) $$\bar{v}_t + (\bar{v} \cdot \mathrm{grad})\bar{v} = -\mathrm{grad}\psi \ .$$

A zero-order solution is still (64) but (65) becomes

(74) $$\nabla^2\psi - \psi_{tt} = -\nabla^2\psi_{tt} \ .$$

The problem of finding an appropriate nonlinear correction to (74) is rather involved. Certainly wherever singularities in ψ appear an equation based on only the quadratic terms would be unsatisfactory.

3. Toward a Bäcklund Transform for the Boussinesq Equation

Hirota's discovery of an N-soliton formula for the Boussinesq equation [34] leads one to conjecture that a Bäcklund transform can also be found. This feeling is supported by the recent publication of inverse scattering operators for (60) by Zakharov [38]. For

(75) $$L \equiv i\frac{\partial^3}{\partial x^3} + i\left(\phi\frac{\partial}{\partial x} + \frac{\partial}{\partial x}\phi\right) + i\frac{\partial}{\partial x} - \frac{1}{\alpha}\Phi_x \ ,$$

(76) $$B \equiv \alpha \frac{\partial^2}{\partial x^2} + \frac{1}{\alpha}\phi ,$$

where $\alpha \equiv \sqrt{3/4}$ and $\phi_t \equiv \Phi_{xx}$, the operator equation

(77) $$iL_t = BL - LB$$

implies (60).

A Bäcklund transform is particularly interesting in this case because the inverse scattering operator L is of third order, and, as Zakharov [38] notes, inverse scattering for a third-order operator "appears difficult and up to now has not been solved." Relations between inverse scattering operators and Bäcklund transforms have been established for several nonlinear wave equations equations by putting the Bäcklund transforms in the form of Riccati equations and interpreting the associated linear equations as the scattering operators [25], [26], [39]. Such a relation may hold even for the third-order equation $L\psi = \lambda\psi$ (with L as in (75))

(78) $$\psi_{xxx} + (2\phi + 1)\psi_x + [\phi_x + \frac{i}{\alpha}(\Phi_x + \lambda\alpha)]\psi = 0$$

if the <u>generalized</u> <u>Riccati</u> <u>equation</u> discussed by Davis [6] is used. Using his notation, the Riccati variable is

(79) $$v \equiv \frac{3}{Q_1}\frac{\psi_x}{\psi} ,$$

but it is not clear what to choose for the function $Q_1(x)$. A hint is obtained by assuming a traveling wave solution of velocity u so

(80) $$\psi_t = -u\psi_x .$$

Then substitution of (79) and (80) into

(81) $$i\psi_t = B\psi ,$$

with the operator B as defined in (76), gives

(82) $$(Q_1 v)_x = -4C - \frac{iu}{\alpha}(Q_1 v) - \frac{1}{3}(Q_1 v)^2$$

where C is a constant level approached by ϕ as $|x| \to \infty$. Since (82) will generate a solitary wave solution for the product $Q_1 v$, it seems reasonable to assume Q_1 is independent of x whereupon the generalized Riccati equation associated with (78) becomes

(83) $$v_{xx} + Q_1 v v_x + (2\phi+1)v + \frac{Q_1^2}{9}v^3 + \frac{3}{Q_1}[\phi_x + \frac{i}{\alpha}(\Phi_x + \lambda\alpha)] = 0$$

and that associated with (81) becomes

(84) $$v_t + i\alpha[v_{xx} + \frac{2}{3} Q_1 v v_x + \frac{4}{Q_1} \phi_x] = 0 \ .$$

Following Wahlquist and Estabrook [39], one can define

(83) $$\phi \equiv -w_x$$

and ask the following question: Does the assumption

(84) $$v = k(w_n - w_{n-1})$$

yield a Bäcklund transform for (60)? This question has been answered affirmatively by Chen [40] and by Hirota [41].

REFERENCES

[1] A.C. SCOTT, Steady propagation on long Josephson junctions, Bull. Amer. Phys. Soc. 12 (1967), 308-309.

[2] ________, A nonlinear Klein-Gordon equation, Amer. J. Phys. 37 (1969), 52-61.

[3] A.C. SCOTT AND W.J. JOHNSON, Internal flux motion in large Josephson junctions, Appl. Phys. Lett. 14 (1969), 316-318.

[4] A.C. SCOTT, Propagation of flux on a long Josephson tunnel junction, Nuovo Cimento B 69 (1970), 241-261.

[5] A. BARONE, F. ESPOSITO, C.J. MAGEE, AND A.C. SCOTT, Theory and applications of the sine-Gordon equation, Riv. Nuovo Cimento 1 (1971), 227-267.

[6] H.T. DAVIS, Introduction to Nonlinear Integral and Differential Equations, Dover Publications, New York, N. Y., 1962, Chap. 3.

[7] D.W. MCLAUGHLIN AND A.C. SCOTT, A restricted Bäcklund transformation, J. Mathematical Phys. 14 (1973), 1817-1828.

[8] P.D. GERBER, Some results on quadratic semilinear differential equations, presented at the Research Workshop on Contact Transformations.

[9] C. ROGERS, Baecklund transformations and invariance properties in axially-symmetric flow, Ann. Soc. Sci. Bruxelles Sér. I 86 (1972), 211-219.

[10] D.S. MOSELEY, Nonseparable solutions of the Helmholtz wave equation, Quart. Appl. Math. 22 (1965), 354-357; Further properties of the nonseparable solutions of the Helmholtz wave equation, ibid. 27 (1969), 451-459.

[11] D.H. SCHOONAERT AND P.J. LUYPAERT, Use of nonseparable solutions of Helmholtz wave equations in waveguides and cavities, Electron. Lett. 9 (1973), 617-618.

[12] A.C. SCOTT, The electrophysics of a nerve fiber, Rev. Modern Phys. 47 (1975), 487-533.

[13] ________, Analysis of nonlinear distributed systems, IRE Trans. Circuit Theory CT-9 (1962), 192-195.

[14] J. NAGUMO, S. ARIMOTO, AND S. YOSHIZAWA, Bistable transmission lines, IEEE Trans. Circuit Theory CT-12 (1965), 400-412.

[15] A.C. SCOTT, Active and Nonlinear Wave Propagation in Electronics, John Wiley and Sons, New York, N. Y., 1970.

[16] G.B. WHITHAM, Linear and Nonlinear Waves, John Wiley and Sons, New York, N. Y., 1974.

[17] L.S. HOEL, W.H. KELLER, J.E. NORDMAN, AND A.C. SCOTT, Niobium superconductive tunnel diode integrated circuit arrays, Solid-State Electron. 15 (1972), 1167-1173.

[18] T.A. FULTON, R.C. DYNES, AND P.W. ANDERSON, The flux shuttle - A Josephson junction shift register employing single flux quanta, Proc. IEEE 61 (1973), 28-35.

[19] J. RUBINSTEIN, Sine-Gordon Equation, J. Mathematical Phys. 11 (1970), 258-266.

[20] A. SEEGER, H. DONTH, AND A. KOCHENDÖRFER, Theorie der Versetzungen in eindimensionalen Atomreihen, Zeit. f. Physik 127 (1950), 533-550; 130 (1951), 321-336; 134 (1953), 173-193.

[21] G.L. LAMB, JR., Analytical descriptions of ultrashort optical pulse propagation in a resonant medium, Rev. Modern Phys. 43 (1971), 99-124.

[22] A.C. SCOTT AND F.Y.F. CHU, Pulse saturation in a traveling wave parametric amplifier, Proc. IEEE 62 (1974), 1720-1721.

[23] A.C. SCOTT, F.Y.F. CHU, AND D.W. MCLAUGHLIN, The soliton: a new concept in applied science, Proc. IEEE 61 (1973), 1443-1483.

[24] A.R. FORSYTH, Theory of Differential Equations, Vol. VI, Dover Publications, New York, N. Y., 1959, p. 454.

[25] M.J. ABLOWITZ, D.J. KAUP, A.C. NEWELL, AND H. SEGUR, The inverse scattering transform-Fourier analysis for nonlinear problems, Studies in Appl. Math. 53 (1974), 249-315.

[26] F.Y.F. CHU AND A.C. SCOTT, Bäcklund transformations and the inverse method, Phys. Lett. 47A (1974), 303-304.

[27] W.J. JOHNSON, Nonlinear wave propagation on superconducting tunneling junctions, Ph.D. Thesis, University of Wisconsin, 1968.

[28] K. NAKAJIMA, T. YAMASHITA, AND Y. ONODERA, Mechanical analog of active Josephson transmission line, J. Appl. Phys. 45 (1974), 3141-3145.

[29] K. NAKAJIMA, Y. ONODERA, T. NAKAMURA, AND R. SATO, Analysis of vortex motions on Josephson line, J. Appl. Phys. 45 (1974), 4095-4099.

[30] R. HIROTA, Exact three-soliton solution of the two-dimensional sine-Gordon equation, J. Phys. Soc. Japan 35 (1973), 1566.

[31] J. BOUSSINESQ, Theorie des ondes et des remous qui se propagent le long d'un canal rectangulaire horizontal, en communiquant au liquide contenu dans ce canal des vitesses sensiblement pareilles de la surface au fond, J. Math. Pures Appl. 7 (1872), 55-108.

[32] N.J. ZABUSKY, A synergetic approach to problems of nonlinear dispersive wave propagation and interaction, in Nonlinear Partial Differential Equations, W.F. Ames, Ed., Academic Press, New York, N. Y., 1967, 223-258.

[33] M. TODA, Studies on a nonlinear lattice, Phys. Rep. 18C (1975), 1-124.

[34] R. HIROTA, Exact N-soliton solution of the wave equation of long waves in shallow-water and in nonlinear lattices, J. Mathematical Phys. 14 (1973), 810-815.

[35] K.E. LONNGREN, H.C.S. HSUAN, AND W.F. AMES, On the soliton, invariant and shock solutions of a fourth-order nonlinear equation, J. Math. Anal. Appl., to be published.

[36] N. HERSHKOWITZ AND T. ROMESSER, Observation of ion-acoustic cylindrical solitons, Phys. Rev. Lett. 32 (1974), 581-583.

[37] S. MAXON AND J. VIECELLI, Spherical solitons, Phys. Rev. Lett. 32 (1974), 4-6.

[38] V.E. ZAKHAROV, On stochastization of one-dimensional chains of nonlinear oscillators, Ž. Èksperimental'noĭ i Teoretičeskoĭ Fiziki 65 (1973), 219-225 (in Russian).

[39] H.D. WAHLQUIST AND F.B. ESTABROOK, Bäcklund transformation for solutions of the Korteweg-deVries equation, Phys. Rev. Lett. 31 (1973), 1386-1390.

[40] H.-H. CHEN, Relation between Bäcklund transformations and inverse scattering problems, this volume.

[41] R. HIROTA, Direct method of finding exact solutions of nonlinear evolution equations, this volume.

ON APPLICATIONS OF GENERALIZED BÄCKLUND TRANSFORMATIONS TO CONTINUUM MECHANICS*

Colin Rogers

University of Western Ontario
London, Ontario, Canada

I. INTRODUCTION

A theory of transformation of surfaces initiated by Bäcklund [1]-[3] and later developed by Loewner [4],[5] has, in recent years, proved to be of remarkable importance in the analysis of a wide range of physical phenomena. In particular, Bäcklund transformations of the sine-Gordon equation have generated results of interest in dislocation theory (Seeger [6]), in the study of long Josephson junctions (Scott [7]), and in the investigation of propagation of long optical pulses through a resonant laser medium (Lamb [8], Barnard [9]). The work by Miura [10] on the Korteweg-deVries equation has likewise involved use of a Bäcklund transformation (see also Whitham [11]). The progress in these areas has been well-documented elsewhere (for example, see Scott, Chu, and McLaughlin [12]) and is beyond the scope of the present survey. However, we note that, in the context of the sine-Gordon equation, the concern is with Bäcklund transformations which leave that equation invariant. Thus a known solution is employed to generate new solutions via a so-called "theorem of permutability." On the other hand, in the cases, for example, of Miura's transformation [10] for the Korteweg-deVries equation and the Hopf-Cole transformation [13], [14] for Burger's equation the Bäcklund transformations are not of an invariance type, but rather, transform the equations to other forms (the modified Korteweg-deVries and linear diffusion equations, respectively). The former leads to the inverse scattering method whereas there is literature available for the latter. In fact, both invariance and reducibility properties under Bäcklund transformations of a generalized nature have also proved of importance in various areas of continuum mechanics. In view

*This work was supported by National Research Council Grant A8780.

of the extent of the work in recent years on this subject only a survey can be presented here.

The invariance and reducibility of the governing equations of gasdynamics and magnetogasdynamics under various transformations has been the subject of numerous papers over the last ten years or so, notably by the Russian School. One may cite the works of Nikol'skii [15], Tomilov [16], Movsesian [17], and Ustinov [18], [19] in gasdynamics and Rykov [20] in magnetogasdynamics. It is noted that Ustinov [19] has utilized invariant transformations to solve certain finite-amplitude shock-wave propagation problems in tubes.

It was Haar [21] who, in connection with a variational problem, first introduced the adjoint transformations which leave invariant (with the exception of the equation of state) the governing equations of plane potential gasdynamics. Later, Bateman [22] constructed other such invariant transformations, namely, the reciprocal relations. The latter were used by Tsien [23] to approximate certain subsonic adiabatic gas flows. Bateman [24] subsequently noted that both the adjoint and reciprocal relations are Bäcklund transformations. Reciprocal properties for steady gasdynamics were further discussed by Power and Smith [25], [26]. These invariant transformations together with Prim's Substitution Principle [27] were placed in the context of a more general formulation of invariant transformations for rotational flow by Rogers [28]; extensions to nondissipative magnetogasdynamics were presented in a recent paper by Rogers, Castell, and Kingston [29]. Reciprocal and adjoint-type invariant transformations for $(\varepsilon+1)$-dimensional spherically symmetric unsteady gasdynamics were established by Rogers [30], [31]. These multi-parameter invariance properties may be applied to generate new solutions from existing solutions of the original system. Application may be made to certain problems involving nonuniform shock-wave propagation (Castell and Rogers [32]). It is interesting to note that whereas there is a significant amount of Russian literature devoted to invariant transformations and their applications in gasdynamics and magnetogasdynamics (see above) this seems to largely have escaped the notice of workers elsewhere. Certainly the method of invariant transformations merits examination in order to see how far it can generate new results of physical significance. Finally, it is mentioned that substitution principles in unsteady gasdynamics and magnetogasdynamics are to be found in papers by Smith [33] and Power and Rogers [34].

In 1950, Loewner [4] introduced an important generalization of the concept of Bäcklund transformation. This was in connection with the reduction to canonical form of the well-known hodograph equations of gasdynamics (specifically, to the Cauchy-Riemann, Tricomi, and wave equations in subsonic, transonic, and supersonic flow, respectively). Such reduction is possible subject to the density-speed relation adopting certain multi-parameter forms. The available parameters may be employed to approximate real-gas behavior over various specified Mach-number ranges (see Power, Rogers, and Osborn [35]). In the latter paper, the unifying aspects of Loewner's work were noted; thus, reductions to canonical form due to Von-Kármán [36], Pérès [37], [38], Coburn [39], Frank'l [40], Kristianovich [41], Sauer [42], Müller [43], and Dombrovskii [44] all emerge naturally from Loewner's formulation. Extensions of the method are available both for certain aligned magnetogasdynamic flows (Rogers [45]) and also for certain gas flows subject to heat addition (Rogers, Kingston, and Castell [46], Castell [47]). A subsequent paper by Loewner [5] introduced the concept of an infinitesimal Bäcklund transformation. However, to the writer's knowledge this theory has yet to be applied to the solution of physical problems.

In fact, Loewner's finite Bäcklund-type transformations may be utilized to advantage in a number of areas of current interest in continuum mechanics outside those of gasdynamics and magnetogasdynamics. In particular, the Stokes-Beltrami equations, being applicable to many classes of physical problems, have received a great deal of attention from authors investigating their various aspects. One may cite, for instance, Weinstein [48] and the interesting work of Ranger [49]. In [48], various applications to such topics as the torsion of shafts of revolution, electrostatics (capacity of conductors), and virtual mass are discussed. In a recent paper by Rogers and Kingston [50] it was shown that generalized Bäcklund transformation may be applied to the Stokes-Beltrami equations with a view to the generation of both invariance and reducibility properties. In a subsequent paper by Clements and Rogers [51] it was shown that the well-known correspondence principle of Weinstein emerges as a particular member of a multi-parameter class of Bäcklund-type transformation of the Stokes-Beltrami equations. Moreover, an iterated form of the correspondence principle was used to obtain solutions to certain boundary-value problems involving axially-symmetric deformations of an incompressible isotropic linear elastic material. Such solutions assume importance in light of recent work by Selvadurai and Spencer [52], where the first-order theory serves as the basis for solutions in second-order incompressible finite elasticity.

Reduction to canonical form of hodograph equations arising in fields other than gasdynamics and magnetogasdynamics has also led to results of interest. In this connection, Rogers and Clements [53] have shown that the hodograph equations of one-dimensional elastic-plastic wave propagation in a medium which exhibits strain-hardening behavior may be reduced to the classical wave equation subject to the stress-strain relation in the plastic region adopting certain multi-parameter forms. One of the latter has already proved useful for the explicit integration of the elastic-plastic wave equations (Courant and Friedrichs [54]). The other forms constructed in [50] may be utilized in a similar manner. An analogous approach may be employed to analyze the propagation of large amplitude waves in such media as saturated soil, dry sand, and clay silt. Thus, multi-parameter stress-strain laws presented recently in an interesting paper by Cekirge and Varley [55] (see also Kazakia and Varley [56]) may be generated via generalized Bäcklund transformations. Moreover, further multi-parameter forms are readily generated via the method (Rogers [57]). Again, the parameters available may be used to approximate real material behavior.

In the above classes of wave propagation problems, there are, of course, the usual complications associated with discussions posed in terms of the hodograph variables. However, for a variety of problems involving wave propagation through various inhomogeneous media, significant progress can be made by use of generalized Bäcklund transformations to seek reduction to a convenient canonical form. Wave propagation in homogeneous elastic media has been widely investigated. However, studies involving wave propagation in inhomogeneous media are much less common; a summary of the literature on this topic is given by Eason [58]. The problem of uniform normal loading of the curved surface of a cylindrically or spherically symmetric inhomogeneous elastic solid was introduced in a subsequent paper by Eason [59] where the possibility of application in geophysical fields was noted. Clements and Rogers [60], [61] have recently applied generalized Bäcklund transformations to the study of wave propagation through elastic media with inhomogeneity. Specific boundary-value problems involving, for example, normal impact loading of a spherical cavity were thereby solved.

The Bäcklund transformation approach may also be used to analyze long wavelength pulse propagation in curved elastic rods (Moodie, Rogers, and Clements [62]-where other methods are indicated) and in the study of (N+1)-dimensional spherically symmetric shear waves in inhomogeneous isotropic viscoelastic media (Rogers, Clements, and Moodie [63]). It is suggested that the exact solutions generated thereby may be used to provide independent verification of the results predicted by the Karal-Keller procedure for obtaining asymptotic wavefront expansions for the travelling disturbance, for example, resulting from dynamic loading of hyperbolic-viscoelastic materials. Again, similar comments apply to the analysis of certain wave propagation problems in fluid-filled elastic tubes (Rogers and Clements [64]). As a further application to Bäcklund transformations to problems involving wave propagation, we mention a recent paper by Clements and Rogers [65] on two-dimensional shallow-water wave generation over curved bottom topographics both with and without ground motion. Current research is being conducted on the application of Bäcklund transformations to the study of radial propagation of rotary shear waves in an incompressible elastic material under finite deformation employing as a basis, Biot's mechanics of incremental deformations (Kurashige [66]).

Reduction to elliptic canonical form via Bäcklund transformations has also been recently used in nonlinear elasto-statics and other areas of continuum mechanics. In the paper by Clements and Rogers [67] application was made to the theory of stress concentration for shear-strained prismatic materials with a nonlinear stress-strain law. Further, a broad class of nonlinear filtration problems are amenable to the technique. Nonlinear filtration is of considerable practical importance, since it arises, in particular, in connection with oil-drilling operations. For example, flows of paraffin-based oils, of oils containing asphaltenes, and of oil-water emulsions are characterized by deviations from linearity in the law relating the rate of filtration and the pressure gradient. Rogers and Swetits [68] constructed a variety of multi-parameter non-Darcian filtration laws for which the hodograph system characterizing the nonlinear filtration is reducible to a form associated with the Cauchy-Riemann equations. As an application of the method, a wide class of nonlinear filtration problems involving a permeable medium with cavities were solved. Finally, it is noted that finite Bäcklund-type transformations may be applied in the study of anti-plane crack problems in inhomogeneous elastic media (Clements, Atkinson, and Rogers [69]).

In summary, it has been seen that the method of Bäcklund-transformations has widespread physical applications. Indeed, whereas an attempt has been made to give an extensive list of recent contributions, the list is by no means exhaustive. It is the contention of the author that there remains much work to be done in exploiting the technique to the fullest extent. The importance of the method is two-fold. Thus, on the one hand, reduction to canonical form of wide classes of physical problems may be made. The existing literature for the canonical form may then be utilized to generate the solution of the original problem. On the other hand, invariant properties may be constructed and employed to generate new solutions from known solutions of nonlinear equations.

II. THEORY

Transformations $\xi_j \to \xi_j^*$, $j = 1,2$, $U \to U^*$ characterized by relations of the type

(2.1) $$B_i(\xi_1,\xi_2,U,U_{\xi_1},U_{\xi_2};\xi_1^*,\xi_2^*,U^*,U^*_{\xi_1^*},U^*_{\xi_2^*}) = 0 ,$$

$$i = 1,\ldots,4 ,$$

where $U_{\xi_j} \equiv \dfrac{\partial U}{\partial \xi_j}$, $U^*_{\xi_j^*} \equiv \dfrac{\partial U^*}{\partial \xi_j^*}$ were introduced by Bäcklund in connection with certain problems involving the transformation of surfaces. In particular, Bäcklund transformations of pseudospherical surfaces which leave the total mean curvature invariant are discussed in Eisenhart [70]. These invariant transformations have important applications in connection with the sine-Gordon equation mentioned earlier. A recent paper by Bauer and Rogers [71] was concerned with Bäcklund transformations and infinitesimal deformation of surfaces.

In analogy with (2.1), generalized Bäcklund-type transformations of the form

(2.2) $$B_i(\xi_1,\xi_2,U_1,\ldots,U_n,U_{1,\xi_1},\ldots,U_{n,\xi_1},U_{1,\xi_2},\ldots,U_{n,\xi_2};$$

$$\xi_1^*,\xi_2^*,U_1^*,\ldots,U_n^*,U^*_{1,\xi_1^*},\ldots,U^*_{n,\xi_1^*},U^*_{1,\xi_2^*},\ldots,U^*_{n,\xi_2^*}) = 0 ,$$

$$i = 1,\ldots,2n+2 ,$$

may be introduced. It is assumed that all functions introduced have continuous derivatives of all occurring orders. It is observed that if ξ_1, ξ_2 are regarded as independent variables, and $U_1,\ldots, U_n$, $U_1^*,\ldots, U_n^*$, ξ_1^*, ξ_2^* are dependent variables, the number of equations in (2.2) is the same as the number of unknowns, namely, 2n+2. Resch [72] has investigated "linear" Bäcklund transformations linking certain systems of partial differential equations involving an arbitrary number of independent variables. Here, attention is confined to the case of linear systems involving two independent variables. Even with these restrictions, the transformations have widespread application. The application of generalized

Bäcklund transformations to systems of nonlinear partial differential equations remains to be studied.

The concern is with the application of certain transformations of the type (2.2) to systems of n linear first-order partial differential equations

$$\sum_{k=1}^{n}[\alpha_{ik}U_{k,\xi_1}+\beta_{ik}U_{k,\xi_2}+\gamma_{ik}U_k]+\delta_i=0,\ i=1,\ldots,n\ , \tag{2.3}$$

for the $U_k(\xi_j)$, $k=1,\ldots,n$, $j=1,2,$. In (2.3), α_{ik}, β_{ik}, γ_{ik} are functions only of the ξ_j. Specifically, if (2.3) is written in the matrix form

$$\sum_{j=1}^{2} L_j\Omega_{\xi_j}+L_3\Omega+L_4=0\ , \tag{2.4}$$

where $L_. \equiv [\alpha_{ik}]$, $L_2\equiv[\beta_{ik}]$, $L_3\equiv[\gamma_{ik}]$, $L_4\equiv[\delta_1,\ldots,\delta_n]^T$, and $\Omega\equiv [U_1,\ldots,U_n]^T$, we investigate the subclass of generalized Bäcklund transformations of type (2.2) defined by

$$\begin{gathered} R_j\Omega^*_{\xi_j}+S_j\Omega_{\xi_j}+T_j\Omega+U_j\Omega^*+V_j=0\ ,\ j=1,2, \\ \xi^*_j=\xi_j\ ,\quad |R_j|\neq 0\ , \end{gathered} \tag{2.5}$$

where R_j, S_j, T_j, U_j are $n\times n$ matrices and V_j are $n\times 1$ column vectors with entries which are arbitrary functions of the ξ_j. In view of the nonsingularity conditions, $|R_j|\neq 0$, (2.5) may be rewritten in the form

$$\begin{gathered} \Omega^*_{\xi^*_j}=A_j\Omega_{\xi_j}+B_j\Omega+C_j\Omega^*+D_j\ ,\quad j=1,2, \\ \xi^*_j=\xi_j\ , \end{gathered} \tag{2.6}$$

where $A_j\equiv -R_j^{-1}S_j$, $B_j\equiv -R_j^{-1}T_j$, $C_j\equiv -R_j^{-1}U_j$, $D_j\equiv -R_j^{-1}V_j$. It is observed that (2.6) provides $2n+2$ relations.

Transformations $\Omega\to\Omega^*$, $\xi_j\to\xi^*_j$ defined by (2.6) are investigated which transform the system (2.4) to an associated system

$$\sum_{j=1}^{2} L^*_j\Omega^*_{\xi^*_j}+L^*_3\Omega^*+L^*_4=0\ ,\quad \Omega^*=[U^*_1,\ldots,U^*_n]^T\ , \tag{2.7}$$

where the L_j^* are functions of $\xi_j^* = \xi_j$.

Imposition of the cross-differentiation conditions

(2.8) $$\Omega_{\xi_1\xi_2} = \Omega_{\xi_2\xi_1} ,$$

(2.9) $$\Omega^*_{\xi_1^*\xi_2^*} = \Omega^*_{\xi_2^*\xi_1^*} ,$$

on the pair of matrix equations of (2.6) shows that

(2.10) $$(A_1-A_2)\Omega_{\xi_1\xi_2} + (A_{1,\xi_2}-B_2-C_2A_1)\Omega_{\xi_1} + (B_1-A_{2,\xi_1}+C_1A_2)\Omega_{\xi_2}$$
$$+ (B_{1,\xi_2}-B_{2,\xi_1}+C_1B_2-C_2B_1) + (C_{1,\xi_2}-C_{2,\xi_1}+C_1C_2-C_2C_1)\Omega^*$$
$$+ C_1D_2 - C_2D_1 + D_{1,\xi_2} - D_{2,\xi_1} = 0 .$$

Hence, if it is required that

$$A_1 = A_2 , \quad C_{1,\xi_2} - C_{2,\xi_1} + C_1C_2 - C_2C_1 = 0 ,$$

(2.11)
$$B_1 - A_{2,\xi_1} + C_1A_2 - (A_{1,\xi_2}-B_2-C_2A_1)L_1^{-1}L_2 = 0 , \quad |L_1| \neq 0$$
$$B_{1,\xi_2} - B_{2,\xi_1} + C_1B_2 - C_2B_1 - (A_{1,\xi_2}-B_2-C_2A_1)L_1^{-1}L_3 = 0 ,$$
$$D_{1,\xi_2} - D_{2,\xi_1} + C_1D_2 - C_2D_1 - (A_{1,\xi_2}-B_2-C_2A_1)L_1^{-1}L_4 = 0 ,$$

a matrix equation of the form (2.4) results. Further, from (2.7), it is seen that

(2.12) $$\sum_{j=1}^{2} L_j^*\Omega^*_{\xi_j^*} + L_3^*\Omega^* + L_4^* =$$
$$L_1^*A_1L_1^{-1}[L_1\Omega_{\xi_1} + L_1A_1^{-1}L_1^{*-1}L_2^*A_2\Omega_{\xi_2} + L_1A_1^{-1}(B_1+L_1^{*-1}L_2B_2)\Omega$$
$$+ L_1A_1^{-1}L_1^{*-1}(L_1^*D_1 + L_2^*D_2+L_4^*)] + (L_1^*C_1+L_2^*C_2+L_3^*)\Omega^*, \quad |L_1^*| \neq 0 ,$$

whence, if we set

$$L_1^{-1}L_2 = -H\ , \quad L_1^{-1}L_3 = -J\ , \quad L_1^{-1}L_4 = -K\ ,$$

(2.13)

$$L_1^{*-1}L_2^* = -H^*\ , \quad L_1^{*-1}L_3^* = -J^*\ , \quad L_1^{*-1}L_4^* = -K^*\ ,$$

then provided

$$H = A_1^{-1}H^*A_2\ , \quad |A_1| \neq 0\ , \quad J = A_1^{-1}(H^*B_2-B_1)\ ,$$

(2.14)

$$J^* = C_1 - H^*C_2\ , \quad K^* = A_1K + D_1 - H^*D_2\ ,$$

the matrix equation (2.12) yields

(2.15) $$\Omega^*_{\xi_1} - H^*\Omega_{\xi_2} - J^*\Omega^* - K^* = A_1(\Omega_{\xi_1} - H\Omega_{\xi_2} - J\Omega - K)\ .$$

Thus, since A_1 is nonsingular, it is concluded that the system (2.4) is transformed to the associated system (2.7) and conversely via the transformations defined by (2.6). It is noted that if the Ω_{ξ_i} are isolated in the latter and the requirement (2.8) is imposed, then no further restrictions arise, since only the condition (2.10) again results.

The following result has therefore been established:

Theorem: *The system*

$$\Omega_{\xi_1} = H\Omega_{\xi_2} + J\Omega + K\ ,$$

is transformed to the associated system

$$\Omega^*_{\xi^*_1} = H^*\Omega^*_{\xi^*_2} + J^*\Omega^* + K^*\ ,$$

and conversely via the generalized Bäcklund-type transformations

$$\Omega^*_{\xi^*_j} = A_j\Omega_{\xi_j} + B_j\Omega + C_j\Omega^* + D_j\ ,$$

$$\xi^*_j = \xi_j\ ,$$

when the following conditions hold:

$$C_{1,\xi_2} - C_{2,\xi_1} + C_1C_2 - C_2C_1 = 0 ,$$

(2.16)

$$B_1 - A_{2,\xi_1} + C_1A_2 + (A_{1,\xi_2} - B_2 - C_2A_1)H = 0 ,$$

$$B_{1,\xi_2} - B_{2,\xi_1} + C_1B_2 - C_2B_1 + (A_{1,\xi_2} - B_2 - C_2A_1)J = 0 ,$$

$$D_{1,\xi_2} - D_{2,\xi_1} + C_1D_2 - C_2D_1 + (A_{1,\xi_2} - B_2 - C_2A_1)K = 0 ,$$

where

(2.17)

$$A_1 = A_2 , \; H = A_1^{-1}H^*A_1 , \; |A_1| \neq 0 ,$$

$$J = A_1^{-1}(H^*B_2 - B_1) , \; J^* = C_1 - H^*C_2 ,$$

$$K^* = A_1K + D_1 - H^*D_2 .$$

The above result includes as a particular case $(n = 2, J = J^* = K = K^* = D_1 = D_2 = 0)$ that obtained by Loewner [4] (see also Bers [73]) in connection with the reduction of the hodograph equations of gasdynamics to canonical form in subsonic, transonic, and supersonic flow.

It is noted that any solution of (2.16) may be represented in the form

(2.18) $$C_j = L_{\xi_j} L^{-1} , \; j = 1,2,$$

where L is a nonsingular $n \times n$ matrix. Accordingly, if matrices $\tilde{A}_j$, $\tilde{B}_j$, $\tilde{D}_j$, $j = 1,2$, are introduced where

(2.19) $$A_j = L\tilde{A}_j , \; B_j = L\tilde{B}_j , \; D_j = L\tilde{D}_j ,$$

the conditions (2.11) reduce to consideration of (2.18) together with

(2.20a) $$\tilde{A}_{2,\xi_1} - \tilde{A}_{1,\xi_2}H - \tilde{B}_1 + \tilde{B}_2 H = 0 \ ,$$

(2.20b) $$\tilde{B}_{1,\xi_2} - \tilde{B}_{2,\xi_1} + (\tilde{A}_{1,\xi_2} - \tilde{B}_2)J = 0 \ ,$$

(2.20c) $$\tilde{D}_{1,\xi_2} - \tilde{D}_{2,\xi_1} + (\tilde{A}_{1,\xi_2} - \tilde{B}_2)K = 0 \ .$$

In many physical contexts, the concern is with cases where $J = J^* = K = K^* = 0$, whence, from (2.20b,c),

(2.21) $$\tilde{B}_j = M_{\xi_j} \ , \ \tilde{D}_j = P_{\xi_j} \ , \ j = 1,2,$$

where M and P are arbitrary matrices. The matrix equation (2.20a) reduces to

(2.22) $$\tilde{A}_{1,\xi_1} - \tilde{A}_{1,\xi_2}H - M_{\xi_1} + M_{\xi_2}H = 0$$

which is linear in $\tilde{A}_1$. A particular solution is

$$\tilde{A}_1 = M \ ,$$

whence, the general solution is

(2.23) $$\tilde{A}_1 = M + N \ ,$$

where N is the general solution of

(2.24) $$N_{\xi_1} - N_{\xi_2}H = 0 \ .$$

Thus, the following result obtains:

Corollary: *The system*

$$\Omega_{\xi_1} = H\Omega_{\xi_2} \ ,$$

is transformed to the associated system

$$\Omega^*_{\xi_1^*} = H^*\Omega^*_{\xi_2^*}$$

and conversely by the generalized Bäcklund-type transformations

$$\Omega^*_{\xi^*_j} = L(M+N)\Omega_{\xi_j} + LM_{\xi_j}\Omega + L_{\xi_j}L^{-1}\Omega^* + LP_{\xi_j} ,$$

$$\xi^*_j = \xi_j , \quad j = 1,2,$$

<u>where</u> L, M, N, <u>and</u> P <u>are</u> $n \times n$ <u>matrices satisfying</u>

$$H = [L(M+N)]^{-1}H^*[L(M+N)] , \quad |L(M+N)| \neq 0 ,$$

$$L_{\xi_1} - H^*L_{\xi_2} = 0 , \quad (LM)_{\xi_1} - H^*(LM)_{\xi_2} = 0 ,$$

$$N_{\xi_1} - N_{\xi_2}H = 0 , \quad P_{\xi_1} - H^*P_{\xi_2} = 0 .$$

III. REDUCTION TO CANONICAL FORMS

Attention is henceforth confined to the class of transformations of the theorem in the preceding section with the specializations $n = 2$, $J = J^* = K = K^* = C_j = D_j = 0$. Moreover, the matrices A_j, B_j are taken to depend only on the variable ξ_1. Hence, the results of the previous section show that

$$\Omega_{\xi_1} = H\Omega_{\xi_2} \leftrightarrow \Omega^*_{\xi^*_1} = H^*\Omega^*_{\xi^*_2} \tag{3.1}$$

via the matrix transformations

$$\begin{aligned} \Omega^*_{\xi_1} &= A_1\Omega_{\xi_1} + H^*B_2\Omega, \quad |A_1| \neq 0 , \\ \Omega^*_{\xi_2} &= A_1\Omega_{\xi_2} + B_2\Omega , \\ \xi^*_i &= \xi_i , \quad i = 1,2 , \end{aligned} \tag{3.2}$$

where B_2 is a constant matrix and

$$A_{1,\xi_1} - H^*B_2 + B_2A_1^{-1}H^*A_1 = 0 , \tag{3.3}$$

$$H = A_1^{-1}H^*A_1 . \tag{3.4}$$

In particular, in applications we are often concerned with systems with

(3.5)
$$H = \begin{bmatrix} 0 & h_2^1 \\ h_1^2 & 0 \end{bmatrix} ,$$

(3.6)
$$H^* = \begin{bmatrix} 0 & h_2^{1*} \\ h_1^{2*} & 0 \end{bmatrix} .$$

With these assumptions, it is necessary to specialize the matrix A_1 so that the property of zero diagonal elements is preserved under the mapping $H \to H^*$. From (3.4), it is clear that this property is invariant if (but not only if) A_1 adopts the diagonal form

(3.7)
$$A_1 = \begin{bmatrix} a_1^1 & 0 \\ 0 & a_2^2 \end{bmatrix} ,$$

in which case

(3.8)
$$H^* = A_1 H A_1^{-1} = \begin{bmatrix} 0 & a_1^1 h_2^1 / a_2^2 \\ a_2^2 h_1^2 / a_1^1 & 0 \end{bmatrix} ,$$

while (3.3) yields

(3.9)
$$a^1_{1,\xi_1} - h_2^{1*} b_1^2 + h_1^{2*} b_2^1 (a_1^1 / a_2^2) = 0 ,$$
$$a^2_{2,\xi_1} - h_1^{2*} b_2^1 + h_2^{1*} b_1^2 (a_2^2 / a_1^1) = 0 ,$$

where

$$B_2 = \begin{bmatrix} 0 & b_2^1 \\ b_1^2 & 0 \end{bmatrix} .$$

Combination of $(3.9)_a$ and $(3.9)_b$ shows that

(3.11)
$$\det A_1 = a_1^1 a_2^2 = \text{constant} = \lambda \neq 0 ,$$

whence the system may be reduced to a single Riccati equation in either a_1^1 or a_2^2. In particular, the Riccati equation in a_1^1 is

$$a^1_{1,\xi_1} + \alpha(a^1_1)^2 + \beta = 0 \ , \quad \text{where} \quad \alpha = h_1^{2*}b_2^1/\lambda, \ \beta = -b_1^2 h_2^{1*} \ . \tag{3.12}$$

Loewner introduced the above class of transformations with a view to the reduction to canonical form of the hodograph equations of gasdynamics for various multi-parameter speed-density relations. The hodograph equations were placed in the matrix form $\Omega_{\xi_1} = H\Omega_{\xi_2}$ and reduction to canonical forms associated with the Cauchy-Riemann, Tricomi, and wave equations was sought for subsonic, transonic, and supersonic flow, respectively. Thus, H* was taken to adopt, in turn, the forms

$$\begin{bmatrix} 0 & -1 \\ 1 & 0 \end{bmatrix}, \begin{bmatrix} 0 & -\xi_1 \\ 1 & 0 \end{bmatrix}, \begin{bmatrix} 0 & 1 \\ 1 & 0 \end{bmatrix} . \tag{3.13}$$

Hence, specifically for the elliptic and hyperbolic cases $(3.13)_a$ and $(3.13)_c$:

(3.14)

(a) If $\beta = 0$,

$$a^1_1 = 1/(\alpha\xi_1+\epsilon).$$

(b) If $\alpha = 0$, then

$$a^1_1 = -\beta\xi_1 + \delta \ .$$

(c) If $\beta/\alpha > 0$, then

$$a^1_1=(\beta/\alpha)^{\frac{1}{2}}\cot\{(\beta/\alpha)^{\frac{1}{2}}(\alpha\xi_1+\zeta)\} \ .$$

(d) If $\beta/\alpha < 0$, then

$$a^1_1=(-\beta/\alpha)^{\frac{1}{2}}\tanh\{(-\beta/\alpha)^{\frac{1}{2}}(\alpha\xi_1+\zeta)\} \ .$$

In fact, as may be seen by consulting the references cited earlier, these reductions to canonical form and their extensions have applications in many areas of continuum mechanics outside gasdynamics. Two specific illustrations are outlined in Section V.

IV. ITERATION AND THE STOKES-BELTRAMI EQUATIONS

Consider the matrix transformations

$$\Omega^*_{\xi_1^*} = A_1\Omega_{\xi_1} + B_1\Omega ,$$

$$(4.1)\qquad \Omega^*_{\xi_2^*} = A_2\Omega_{\xi_2} + B_2\Omega ,$$

$$\xi_j^* = \xi_j , \quad j = 1,2 ,$$

$$\Omega = [U_1,U_2]^T , \quad \Omega^* = [U_1^*,U_2^*]^T ,$$

with

$$(4.2)\qquad A_1 = A_2 = \begin{bmatrix} 0 & \xi_1^{-p-1} \\ -\xi_1^{p+1} & 0 \end{bmatrix} ,$$

$$(4.3)\qquad B_1 = \begin{bmatrix} 0 & 0 \\ -(p+1)\xi_1^p & 0 \end{bmatrix} , \quad B_2 = \begin{bmatrix} -(p+1) & 0 \\ 0 & 0 \end{bmatrix} .$$

These transformations link the Stokes-Beltrami system

$$(4.4)\qquad \Omega_{\xi_1} = H\Omega_{\xi_2} , \quad H = \begin{bmatrix} 0 & -\xi_1^{-p-2} \\ \xi_1^{p+2} & 0 \end{bmatrix} ,$$

with the associated system

$$(4.5)\qquad \Omega^*_{\xi_1^*} = H^*\Omega^*_{\xi_2^*} , \quad H^* = \begin{bmatrix} 0 & -\xi_1^{-p} \\ \xi_1^p & 0 \end{bmatrix} .$$

Explicitly, these transformations yield

$$U^*_{1,\xi_1} = \xi_1^{-p-1}U_{2,\xi_1} , \quad U^*_{1,\xi_2} = \xi_1^{-p-1}U_{2,\xi_2} - (p+1)U_1 ,$$

$$(4.6)\qquad U_2^* = -\xi_1^{p+1}U_1 .$$

The latter relation is Weinstein's correspondence principle

$$(4.7)\qquad U_2\{p\} = C\xi_1^{p+1}U_1\{p+2\}$$

with $C = -1$. Iteration may be used to solve certain boundary-value problems involving axially symmetric deformation due to solid inclusions in certain rubber-like materials (Clements and Rogers [51]). It is readily established by iteration of the above transformation that the general solution of the system (4.5) for $p = 2N$, $N = 0,1,\ldots$ is given by

$$U_1^* + i\xi_1^{-2N}U_2^* = \sum_{r=0}^{N} \frac{(-1)^r 2^r (2N-r)!\,\xi_1^{-2N+r}}{r!\,(N-r)!} \left\{ \frac{(N-r)\,\Phi^{(r)}(\zeta)}{(2N-r)} - \frac{N}{(2N-r)}\overline{\Phi^{(r)}(\zeta)} \right\} \tag{4.8}$$

$$\equiv \Lambda(\xi_1,\xi_2)\ , \left(\frac{N}{2N-r} \overset{\text{def}}{=} 1\ ,\ \frac{N-r}{2N-r} \overset{\text{def}}{=} 0\ , \text{ when } r = N = 0 \right)$$

where $\Phi^{(r)}(\zeta) \equiv \dfrac{\partial^r}{\partial\zeta^r}\{\Phi(\zeta)\}$, $\phi(\zeta)$ being an arbitrary analytic function of $\zeta \equiv \xi_1 + i\xi_2$ and $\overline{\Phi^{(r)}(\zeta)}$ the complex conjugate of $\Phi^{(r)}(\zeta)$.

Further, the transformation

$$U_1^* \to U_2^\dagger\ ,\ U_2^* \to -U_1^\dagger\ ,\ \xi_j^\dagger = \xi_j^*\ , \tag{4.9}$$

takes the matrix system (4.5) to

$$\Omega^\dagger_{\xi_1^\dagger} = H^\dagger \Omega^\dagger_{\xi_2^\dagger}\ ,\ \Omega^\dagger = [U_1^\dagger, U_2^\dagger]^T \tag{4.10}$$

where

$$H^\dagger = \begin{bmatrix} 0 & -\xi_1^p \\ \xi_1^{-p} & 0 \end{bmatrix}. \tag{4.11}$$

Hence, the solution of the Stokes-Beltrami system defined by (4.10) and (4.11) with $p = 2N$, $N = 0,1,\ldots$ is given by

$$U_2^\dagger - iU_1^\dagger = \Lambda(\xi_1,\xi_2) \; . \tag{4.12}$$

In a similar fashion it may be shown that the general solution of the hyperbolic system

$$\Omega^*_{\xi_1} = H^*\Omega^*_{\xi_2} \; , \quad H^* = \begin{bmatrix} 0 & \xi_1^{-p} \\ \xi_1^p & 0 \end{bmatrix} \tag{4.13}$$

for $p = 2N$, $N = 0,1,\ldots$ is given by

$$U_1^* = \sum_{r=0}^{N} \frac{(-1)^r 2^r (2N-r)!\,\xi_1^{-2N+r}}{r!(N-r)!(2N-r)} \,\, {}_r\,[F(\xi_2+\xi_1) + (-1)^r G(\xi_2-\xi_1)] \tag{4.14}$$

$$\equiv f(\xi_1,\xi_2) \; ,$$

$$U_2^* = \sum_{r=0}^{N} \frac{(-1)^r 2^r (2N-r)!\,\xi_1^r}{r!(N-r)!} [F^{(r+1)}(\xi_2+\xi_1) - (-1)^r G^{(r+1)}(\xi_2-\xi_1)] \tag{4.15}$$

$$\equiv g(\xi_1,\xi_2) \; ,$$

where F, G are arbitrary functions of $\xi_2+\xi_1$ and $\xi_2-\xi_1$, respectively, while $\Phi^{(r)}(\zeta)$ denotes the rth derivative of Φ with respect to its indicated argument ζ. The case $N = 0$ provides the familiar d'Alembert solution, while the solution for $N = 1$ has been employed by Clements and Rogers [60] to solve a class of initial-value problems related to wave propagation through inhomogeneous elastic media.

Again, a simple transformation

$$U_1^* \to U_2^\dagger \; , \quad U_2^* \to U_1^\dagger \; , \quad \xi_j^\dagger = \xi_j^* \tag{4.16}$$

shows that the solution of the system (4.13) for $p = -2N$, $N = 0,1,\ldots$ is given by

$$U_1^* = g(\xi_1,\xi_2) \; , \quad U_2^* = f(\xi_1,\xi_2) \; . \tag{4.17}$$

V. APPLICATION

As indicated in the Introduction, there is a wide variety of areas in continuum mechanics where quite simple Bäcklund-type transformations may be of value either in the generation of invariance properties or in the reduction of problems to canonical forms amenable to solution by classical techniques. In some cases the boundary conditions in the transformed problem may be considerably complicated by the procedure. However this is by no means always the case (for example, see [60]). In many cases, use is made of a hodograph transformation before reduction is sought via Bäcklund transformations. There are then, in general, the usual difficulties associated with boundary-value problems transformed to the hodograph plane. In other cases no hodograph transformation is necessary to achieve reduction. Examples of both kinds of situation are indicated below.

1. Aligned Nondissipative Magnetogasdynamics

The governing equations of the steady flow of a thermally nonconducting Prim gas of infinite electrical conductivity in the presence of a magnetic field are (in rationalized MKS units)

(5.1) $$u_j \rho_{,j} + \rho u_{j,j} = 0 \;,$$

(5.2) $$\rho u_j u_{i,j} + \Pi_{,i} - \mu H_j H_{i,j} = 0 \;, \quad i = 1,2,3, \;,$$

(5.3) $$H_{j,j} = 0 \;,$$

(5.4) $$u_{i,j} H_j - H_{i,j} u_j - H_i u_{j,j} = 0 \;, \quad i = 1,2,3$$

(5.5) $$u_j [S(s)]_{,j} = 0 \;,$$

(5.6) $$\rho = P(p) S(s) \;,$$

where $\Pi = p + \frac{1}{2}\mu H_j H_j$, μ is the magnetic permeability (here assumed constant), while u_i and H_i designate the rectangular Cartesian components of the velocity and magnetic fields, respectively. Further, ρ, p, and s denote the gas density, gas pressure, and specific entropy, respectively. It is assumed that there are no external forces present. To (5.1)-(5.6) are adjoined appropriate boundary conditions (for example, see Rogers and Kingston [74]). Thus, for instance, for the super-Alfvénic aligned conducting flow past cylindrical insulars for which the present analysis is particularly straightforward, the discontinuity in the tangential component of the magnetic field implies the existence of a surface current in the fluid layer adjacent to the insulator (Grad [75], Stewartson [76]).

It is readily shown (Iurev [77]) that the equations of homentropic, plane, nondissipative magnetogasdynamics with aligned velocity and magnetic fields may be reduced, under the additional assumption of uniform stagnation enthalpy to the gasdynamic-type system

$$u_j^* \rho_{,j}^* + \rho^* u_{j,j}^* = 0 \quad , \tag{5.7}$$

$$\rho^* u_j^* u_{i,j}^* + p_{,i}^* = 0 \quad , \quad i = 1,2 \; , \tag{5.8}$$

where the variables u_i^*, ρ^*, p^* are given by

$$\begin{gathered} p^* \equiv \Pi \;, \; q^* \equiv q(1-\mu k^2\rho) \;, \\ \rho^* \equiv \rho(1-\mu k^2\rho)^{-1} \;, \; 1 - \mu k^2\rho \neq 0 \;, \\ q^* e^{i\theta} \equiv u^* + iv^* \;, \quad q e^{i\theta} \equiv u + iv \;, \end{gathered} \tag{5.9}$$

and u_j, u_j^*, $j = 1,2$ are denoted in turn by u, v and u^*, v^*. It is assumed that the k occurring in the aligned condition

$$H_i = k\rho u_i, \; , \quad i = 1,2 \; , \tag{5.10}$$

is a constant.

Stream and potential functions ψ and ϕ may now be introduced according to

$$\nabla\Omega = \begin{bmatrix} u^* & v^* \\ -\rho^* v^* & \rho^* u^* \end{bmatrix} \nabla\zeta \tag{5.11}$$

where

(5.12) $$\Omega = \begin{bmatrix} \phi \\ \psi \end{bmatrix}, \quad \zeta = \begin{bmatrix} x \\ y \end{bmatrix}.$$

A hodograph system

(5.13) $$\Omega_{q*} = \Lambda\Omega_\theta$$

where

(5.14) $$\Lambda = \begin{bmatrix} 0 & \dfrac{(M^2-1)(1-\mu k^2\rho)}{[1-\mu k^2\rho(1-M^2)]\rho q} \\ \dfrac{\rho}{(1-\mu k^2\rho)^2 q} & 0 \end{bmatrix},$$

(5.15) $$0 < |J(q*,\theta,x,y)| < \infty,$$

is now established in the usual manner; in (5.14), M is the local Mach number.

If we now set

(5.16) $$s = \pm\int_{q_0}^{q}\left[\frac{(1-M^2)(1-\mu k^2\rho(1-M^2))}{1-\mu k^2}\right]^{1/2}\frac{dq}{q}, \quad t = \theta$$

(where the negative sign is taken for the interval of variation of q in which $dq*/ds < 0$) and q_0 is a reference speed, the system (5.13) reduces to (if s and $K^{1/2}$ are real)

(5.17) $$\Omega_s = \Lambda^*\Omega_t$$

where

(5.18) $$\Lambda^* = \begin{bmatrix} 0 & -K(s)^{1/2} \\ K(s)^{-1/2} & 0 \end{bmatrix}$$

and $K(s)$ is defined by

(5.19) $$K(s) = \frac{(1-M^2)(1-\mu k^2\rho)^3}{\rho^2(1-\mu k^2\rho(1-M^2))}.$$

If s and $K^{1/2}$ are imaginary, the hodograph equations reduce to the hyperbolic form

(5.20)
$$\Omega_\sigma = \Lambda^{**}\Omega_\tau$$

where

(5.21)
$$\Lambda^{**} = \begin{bmatrix} 0 & \chi^{1/2} \\ \chi^{-1/2} & 0 \end{bmatrix}$$

and

(5.22)
$$\sigma = -is , \quad \chi^{1/2} = -iK^{1/2} .$$

In sub-Alfvénic regions $(\mu k^2\rho > 1)$, it is seen that, if $M > 1$, the hodograph system is elliptic. If $M < 1$, since $(\mu k^2\rho)^{-1} = A^2 = q^2/b^2$ where $b = (\mu/\rho)^{1/2}H$ is the Alfvén speed it is seen that the system is elliptic in the region

$$\frac{q^2}{a^2} + \frac{q^2}{b^2} < 1$$

and hyperbolic in the region

$$\frac{q^2}{a^2} + \frac{q^2}{b^2} > 1 .$$

In super-Alfvénic regions $(\mu k^2\rho < 1)$, as in conventional gasdynamics, the hodograph system is elliptic for $M < 1$ and hyperbolic for $M > 1$.

It is observed that the elliptic and hyperbolic matrix equations (5.17), (5.18) are in a form suitable for the application of matrix Bäcklund-type transformations and the results obtained earlier apply. In particular, if attention is restricted to super-Alfvénic subsonic flow and we introduce the simple Bäcklund transformation

(5.23)
$$\begin{bmatrix} \phi' \\ \psi' \end{bmatrix} = \begin{bmatrix} c^{-1} & 0 \\ 0 & c \end{bmatrix} \begin{bmatrix} \phi \\ \psi \end{bmatrix} , \quad s' = s, \; t' = t ,$$

where c is a real nonzero constant, then the new hodograph matrix H' is given by $(A > 1, M < 1)$

$$H' = \begin{bmatrix} 0 & K^{1/2}c^{-2} \\ -c^2K^{-1/2} & 0 \end{bmatrix} .$$

Hence, subject to K being approximated by an expression of the type

$$K^{\dagger} = c^4 \tag{5.24}$$

the magnetogasdynamic system defined by (5.13), (5.14) is transformed via (5.23) to the Cauchy-Riemann equations

$$\phi'_s = -\psi'_\theta \ , \ \psi'_s = \phi'_\theta \ . \tag{5.25}$$

It is readily shown that the approximated q, ρ, p corresponding to the approximation (5.24) are given by

$$\begin{aligned} q^{\dagger} &= d[\operatorname{cosech}(s_0-s)+\mu k^2c^{-2}\operatorname{sech}(s_0-s)] \ , \\ \rho^{\dagger} &= \frac{c^{-2}\tanh(s_0-s)}{1+\mu k^2c^{-2}\tanh(s_0-s)} \ , \\ p^{\dagger} &= -c^2d^2[1+\mu k^2c^{-2}\tanh(s_0-s)]\coth(s_0-s) + \\ &\quad + \mu k^2c^{-4}d^2 - \frac{1}{2}\mu k^2c^{-4}d^2\operatorname{sech}^2(s_0-s) + e \ , \end{aligned} \tag{5.26}$$

where c, d, e, and s_0 are constants available for approximation purposes. The latter relation can be written as

$$p^{\dagger} = -c^4d^2\rho^{\dagger -1} + \frac{\mu k^2d^2\rho^{\dagger 2}}{2(1-\mu k^2\rho^{\dagger})^2} + \frac{\mu k^2c^{-4}d^2}{2} + e \ , \tag{5.27}$$

which is the approximated "equation of state" in the aligned magnetogasdynamic flow. In the limit $K \to 0$, the celebrated Kármán-Tsien approximation of non-conducting gasdynamics is recovered.

2. Wave Propagation in Inhomogeneous Elastic Media with $(\varepsilon+1)$-Dimensional Spherical Symmetry

The appropriate equation of motion in the case of uniform normal loading of an $(\varepsilon+1)$-dimensional spherically symmetric surface $(\varepsilon = 0,1,2)$ is

$$(5.28)\qquad \frac{\partial\sigma_r}{\partial r} + \varepsilon(\sigma_r - \sigma_\theta)r^{-1} = \rho\frac{\partial^2 u}{\partial t^2} ,$$

while the stress-strain relations adopt the forms

$$(5.29)\qquad \sigma_r = \xi\frac{\partial u}{\partial r} + \varepsilon\lambda u r^{-1} , \quad \xi = \lambda + 2\mu ,$$

$$(5.30)\qquad \sigma_\theta = \lambda\frac{\partial u}{\partial r} + (\varepsilon\lambda + 2\mu)ur^{-1} , \quad \varepsilon = 1,2 .$$

The case $\varepsilon = 0$ refers to longitudinal wave propagation in an inhomogeneous slab, λ and μ being the Lamé "constants"; for longitudinal wave propagation in an inhomogeneous rod, $\xi = E$ where E is Young's modulus. In the latter context, r denotes the spatial coordinate measured along the rod, σ_r is the stress, and u is the longitudinal displacement. For $\varepsilon = 1$, the equation of motion (5.28) is for an inhomogeneous elastic solid subjected to uniform normal leading over a cylindrical surface; in that case, u is the radial displacement, and σ_r, σ_θ are the radial and circumferential stress components in the usual cylindrical (r,θ,z)-coordinates. The relations (5.29) and (5.30) correspond to conditions of plane strain; ρ, λ, and μ are assumed to be dependent on r alone. Finally, for $\varepsilon = 2$ the situation described corresponds to that of uniform normal loading of a spherical surface so that the only nonzero displacement is the radial displacement in the radial component u. Just as for the cases $\varepsilon = 0,1$, the quantities ρ, λ, and μ are taken to be dependent on the radial coordinate r alone.

Combination of (5.28)-(5.30) provides an equation of motion in the form

$$(5.31)\qquad \xi\left[\frac{\partial^2 u}{\partial r^2} + \frac{\varepsilon}{r}\frac{\partial u}{\partial r} - \frac{\varepsilon u}{r^2}\right] + \xi'\left[\frac{\partial u}{\partial r} + \frac{\varepsilon u}{r}\right] - \frac{2\varepsilon u}{r}\mu' = \rho\frac{\partial^2 u}{\partial t^2}$$

where the primed quantities refer to derivatives with respect to r. It is convenient to introduce a function $w(r,t)$ so that the governing equation (5.31) emerges as a consequence of the matrix system

$$(5.32)\qquad \Omega_r = M\Omega_t + N\Omega ,$$

where

(5.33) $$\Omega = \begin{bmatrix} u \\ w \end{bmatrix}, \quad M = \begin{bmatrix} 0 & m_{12} \\ m_{21} & 0 \end{bmatrix}, \quad N = \begin{bmatrix} n_{11} & 0 \\ 0 & 0 \end{bmatrix}$$

and m_{12}, m_{21}, n_{11} are dependent on r alone. The requirement that elimination of w produces (5.31) yields

(5.34) $$\det M = -\rho/\xi \ ,$$
$$\xi r^{\varepsilon} m_{12} = \exp\{-\int n_{11} dr\} \ ,$$
$$\xi m_{12} [n_{11}/m_{12}]' = \varepsilon\{\tfrac{2}{r} \mu' - (\xi/r)'\}.$$

If we set

$$\Phi \equiv \exp\{-\int n_{11} dr\}$$

the system (5.34) gives

(5.35) $$m_{12} = \xi^{-1} r^{-\varepsilon} \Phi \ , \ m_{21} = \rho r^{\varepsilon}/\Phi \ , \ n_{11} = -\Phi'/\Phi \ ,$$

while Φ satisfies the equation

(5.36) $$r^{-\varepsilon}\Phi[r^{\varepsilon}\xi\Phi'/\Phi^2]' = \varepsilon\{(\xi/r)' - (2/r)\mu'\} \ .$$

Introduction of the new independent variables

(5.37) $$u^* = \Phi u \ , \quad w^* = u \ ,$$

reduces the system (5.32) to

(5.38) $$\begin{bmatrix} u^* \\ w^* \end{bmatrix}_r = \begin{bmatrix} 0 & \xi^{-1} r^{-\varepsilon} \Phi^2 \\ \rho r^{\varepsilon}/\Phi^2 & 0 \end{bmatrix} \begin{bmatrix} u^* \\ w^* \end{bmatrix}_t \ .$$

If new independent variables r^*, t^* are now introduced according to

(5.39) $$r^* = \int_{r_0}^{r} (\rho/\xi)^{1/2} \, dR \ , \quad t^* = t \ ,$$

the system (5.38) reduces to

$$\begin{bmatrix} u^* \\ w^* \end{bmatrix}_{r^*} = \begin{bmatrix} 0 & K^{-1/2} \\ K^{1/2} & 0 \end{bmatrix} \begin{bmatrix} u^* \\ w^* \end{bmatrix}_{t^*} , \tag{5.40}$$

where

$$K = \rho \xi r^{2\epsilon} \Phi^{-4} . \tag{5.41}$$

Matrix Bäcklund-type transformations may now be applied to the system (5.40) in the usual manner and reduction achieved to a form associated with the conventional wave equation for various multi-parameter forms for ρ and ξ. The constants available in the latter may be utilized to fit ρ and ξ to curves giving their variation for various inhomogeneous elastic materials. Reference may be made to the recent paper by Clements and Rogers [61] where application is made to a specific problem involving the propagation of radially symmetric waves from an $(\epsilon+1)$-dimensional spherical cavity in an infinite inhomogeneous medium. Application of a correspondence principle allows extension of such results to the dynamics of initially dead inhomogeneous viscoelastic media.

VI. CONCLUSION

A survey of applications of Bäcklund-type transformations in continuum mechanics has been presented and some of the recent developments noted. Current work (Clements, Atkinson, and Rogers [69], Rogers, Cekirge, and Askar [78], Clements and Rogers [79]) indicates a link with the linear integral operator technique of Bergman [80]. Thus, the transformations are associated with cases when the series expansions of the latter method terminate. However, in view of its straightforward nature, the value of the Bäcklund transformation approach remains.

REFERENCES

[1] A.V. BÄCKLUND, Ueber Flächentransformationen, Math. Ann. 9 (1876), 297-320.

[2] ____________, Zur theorie der partiellen Differentialgleichung erster ordnung, Math. Ann. 17 (1880), 285-328.

[3]. ____________, Zur theorie der Flächentransformationen, Math. Ann 19 (1882), 387-422.

[4] C. LOEWNER, A transformation theory of partial differential equations of gasdynamics, NACA Tech. Note 2065, 1950.

[5] __________, Generation of solutions of systems of partial differential equations by composition of infinitesimal Baecklund transformations, J. Analyse Math. 2 (1952), 219-242.

[6] A. SEEGER, Theorie der Gitterfehlstellen, Handbuch der Physik 7, no. 1, 383-665, Springer, Berlin, 1955.

[7] A.C. SCOTT, Propagation of flux on a long Josephson tunnel junction, Nuovo Cimento B 69 (1970), 241-261.

[8] G.L. LAMB, JR., Analytical descriptions of ultrashort optical pulse propagation in a resonant medium, Rev. Modern Phys. 43 (1971), 99-124.

[9] T.W. BARNARD, $2N\pi$ Ultrashort light pulses, Phys. Rev. A 7 (1973), 373-376.

[10] R.M. MIURA, Korteweg-deVries equation and generalizations. I. A remarkable explicit nonlinear transformation, J. Math. Phys. 9 (1968), 1202-1204.

[11] G.B. WHITHAM, Linear and Nonlinear Waves, John Wiley and Sons, New York, N.Y., 1974.

[12] A.C. SCOTT, F.Y.F. CHU, AND D.W. MCLAUGHLIN, The soliton: a new concept in applied science, Proc. IEEE 61 (1973), 1443-1483.

[13] E. HOPF, The partial differential equation $u_t + uu_x = \mu u_{xx}$, Comm. Pure Appl. Math. 3 (1950), 201-230.

[14] J.D. COLE, On a quasi-linear parabolic equation occurring in aerodynamics, Quart. Appl. Math. 9 (1951), 225-236.

[15] A.A. NIKOL'SKII, Invariant transformation of the equations of motion of an ideal monatomic gas and new classes of their exact solutions, Prikl. Mat. Meh. 27 (1963), 740-756.

[16] E.D. TOMILOV, On the method of invariant transformations of the gasdynamics equations, Prikl. Mat. Meh. 29 (1965), 959-960.

[17] L.A. MOVSESIAN, On an invariant transformation of equations of one-dimensional unsteady motion of an ideal compressible fluid, Prikl. Mat. Meh. 31 (1967), 137-141.

[18] M.D. USTINOV, Transformation and some solutions of the equation of motion of an ideal gas, Izv. Akad. Nauk SSSR Ser. Meh. Zidk. Gaza 3 (1966), 68-74.

[19] ___________, Ideal gas flow behind an infinite amplitude shock wave, Izv. Akad. Nauk SSSR Ser. Meh. Zidk. Gaza 4 (1967), 88-90.

[20] V.A. RYKOV, On an exact solution of the equations of magnetogasdynamics of finite conductivity, Prikl. Mat. Meh. 29 (1965), 178-181.

[21] A. HAAR, Über adjungierte Variationsprobleme und adjungierte Extremalflächen, Math. Ann. 100 (1928), 481-502.

[22] H. BATEMAN, The lift and drag functions for an elastic fluid in two-dimensional irrotational flow, Proc. Nat. Acad. Sci. U.S.A. 24 (1938), 246-251.

[23] H.S. TSIEN, Two-dimensional subsonic flow of compressible fluids, J. Aero.

Sci. 6 (1939), 399-407.

[24] H. BATEMAN, The transformation of partial differential equations, Quart. Appl. Math. 1 (1943-44), 281-295.

[25] G. POWER AND P. SMITH, Application of a reciprocal property to subsonic flow, Appl. Sci. Res. A8 (1959), 386-392.

[26] ____________________, Reciprocal properties of plane gas flows, J. Math. Mech. 10 (1961), 349-361.

[27] R.C. PRIM, Steady rotational flow of ideal gases, J. Rat. Mech. Anal. 1 (1952), 425-497.

[28] C. ROGERS, The construction of invariant transformations in plane rotational gasdynamics, Arch. Rational Mech. Anal. 47 (1972), 36-46.

[29] C. ROGERS, S.P. CASTELL, AND J.G. KINGSTON, On invariance properties of conservation laws in non-dissipative planar magneto-gasdynamics, J. Mécanique 13 (1974), 343-354.

[30] C. ROGERS, Reciprocal relations in non-steady one-dimensional gasdynamics, Z. Angew. Math. Phys. 19 (1968), 58-63.

[31] _________, Invariant transformations in non-steady gasdynamics and magneto-gasdynamics, Z. Angew. Math. Phys. 20 (1969), 370-382.

[32] S.P. CASTELL AND C. ROGERS, Application of invariant transformations in one-dimensional non-steady gasdynamics, Quart. Appl. Math. 32 (1974), 241-251.

[33] P. SMITH, An extension of the substitution principle to certain unsteady gas flows, Arch. Rational Mech. Anal. 15 (1964), 147-153.

[34] G. POWER AND C. ROGERS, Substitution principles in non-steady magneto-gasdynamics, Appl. Sci. Res. 21 (1969), 176-184.

[35] G. POWER, C. ROGERS, AND R.A. OSBORN, Baecklund and generalised Legendre transformations in gasdynamics, Z. Angew. Math. Mech. 49 (1969), 333-340.

[36] T. VON KÁRMÁN, Compressibility effects in aerodynamics, J. Aero. Sci. 8 (1941), 337-356.

[37] J. PÉRÈS, Quelques transformations des équations du mouvement d'un fluide compressible, Comptes Rendus Acad. Sciences Paris 219 (1944), 501-504.

[38] ________, Sur l'integration des équations qui regissent le mouvement d'un fluide compressible, Proc. 7th Int. Cong. Appl. Mech. 2 (1948), 382-387.

[39] N. COBURN, The Kármán-Tsien pressure-volume relation in the two-dimensional supersonic flow of compressible fluids, Quart. Appl. Math. 3 (1945), 106-116.

[40] F.I. FRANKL', On Chaplygin's problem for mixed sub- and supersonic flows, Izv. Akad. Nauk. SSSR 9 (1945), 121-143.

[41] S.A. KHRISTIANOVICH, On supersonic gas flows, Trudy TSAGI No. 543, 1941.

[42] R. SAUER, Unterschallströmungen um Profile bei quadratisch approximierter Adiabate, Bayer. Akad. Wiss. Math.-Natur. Kl. S.-B. 9 (1951), 65-71.

[43] W. MÜLLER, Gasströmungen bei quadratisch angenäherter Adiabate. Diss. Th. München, 1953.

[44] G.A. DOMBROVSKII, Approximation methods in the theory of plane adiabatic gas flow, Moscow, 1964.

[45] C. ROGERS, Application of Bäcklund transformations in aligned magneto-gasdynamics, Acta Physica Austriaca 31 (1970), 80-88.

[46] C. ROGERS, J.G. KINGSTON, AND S.P. CASTELL, The reduction to canonical form of hodograph equations in elliptic diabatic flow, Acta Physica Austriaca 34 (1971), 242-250.

[47] S.P. CASTELL, Approximate solutions and applications of hodograph equations in elliptic diabatic flow, Acta Physica Austriaca 37 (1973), 193-204.

[48] A. WEINSTEIN, Generalised axially symmetric potential theory, Bull. Amer. Math. Soc. 59 (1953), 20-38.

[49] K.B. RANGER, Some integral transformation formulae for the Stokes-Beltrami equations, J. Math. Mech. 12 (1963), 663-673.

[50] C. ROGERS AND J.G. KINGSTON, Application of Baecklund transformations to the Stokes-Beltrami equations, J. Austral. Math. Soc. 15 (1973), 179-189.

[51] D.L. CLEMENTS AND C. ROGERS, On the application of a Baecklund transformation to linear isotropic elasticity, J. Inst. Math. Appl. 14 (1974), 23-30.

[52] A.P.S. SELVADURAI AND A.J.M. SPENCER, Second order elasticity with axial symmetry - I General Theory, Internat. J. Engrg. Sci. 10 (1972), 97-114.

[53] C. ROGERS AND D.L. CLEMENTS, On the reduction of the hodograph equations for one-dimensional elastic-plastic wave propagation, Quart. Appl. Math. 32 (1975), 469-474.

[54] R. COURANT AND K.O. FRIEDRICHS, Supersonic Flow and Shock Waves, Interscience, New York, N.Y., 1948.

[55] H.M. CEKIRGE AND E. VARLEY, Large amplitude waves in bounded media. I. Reflexion and transmission of large amplitude shockless pulses at an interface, Philos. Trans. Roy. Soc. London Ser. A 273 (1973), 261-313.

[56] J.Y. KAZAKIA AND E. VARLEY, Large amplitude waves in bounded media. II. The deformation of an impulsively loaded slab: the first reflexion; III. The deformation of an impulsively loaded slab: the second reflexion, Philos. Trans. Roy. Soc. London Ser. A 277 (1974), 191-250.

[57] C. ROGERS, Iterated Baecklund-type transformations and the propagation of disturbances in non-linear elastic materials, J. Math. Anal. Appl. 49 (1975), 638-648.

[58] G. EASON, Wave propagation in inhomogeneous elastic media, Bull. Seism. Soc. Amer. 57 (1967), 1267-1277.

[59] ________, Wave propagation in inhomogeneous elastic media; normal leading of spherical and cylindrical surfaces, Appl. Sci. Res. 21 (1970), 467-477.

[60] D.L. CLEMENTS AND C. ROGERS, On wave propagation in inhomogeneous elastic media, Internat. J. Solids and Structures 10 (1974), 661-669.

[61] ________________________, Wave propagation in (N+1)-dimensional spherically symmetric inhomogeneous elastic media, Canad. J. Phys. 52 (1974), 1246-1252.

[62] T. BRYANT MOODIE, C. ROGERS, AND D.L. CLEMENTS, Large wave-length pulse propagation in curved elastic rods, to appear J. Acoustical Soc. Amer.

[63] C. ROGERS, T. BRYANT MOODIE, AND D.L. CLEMENTS, Les ondes de cisaillement à symétrie sphérique en (I+1)-dimensions pour un matériel viscoélastique inhomogène et isotropique, submitted for publication.

[64] C. ROGERS AND D.L. CLEMENTS, Wave propagation in fluid filled elastic tubes, Acta. Mech. 22 (1975), 1-9.

[65] D.L. CLEMENTS AND C. ROGERS, Analytic solution of the linearized shallow-water wave equations for certain continuous depth variations, to appear J. Austral. Math. Soc.

[66] M. KURASHIGE, Radial propagation of rotary shear waves in a finitely deformed elastic solid, Internat. J. Engrg. Sci. 12 (1974), 585-596.

[67] D.L. CLEMENTS AND C. ROGERS, On the theory of stress concentration for shear strained prismatical bodies with a non-linear stress-strain law, Mathematika 22 (1975), 34-42.

[68] C. ROGERS AND J. SWETITS, On a class of non-linear filtration laws, to appear Rheologica Acta.

[69] D.L. CLEMENTS, C. ATKINSON, AND C. ROGERS, Anti-plane crack problems for an inhomogeneous elastic material, submitted for publication.

[70] L.P. EISENHART, A Treatise on the Differential Geometry of Curves and Surfaces, Dover Publications, New York, N.Y., 1960.

[71] K.W. BAUER AND C. ROGERS, Zur infinitesimalen Deformation von Flachen, Mathematische-Statistiche Sektion Forschungszentrum Graz 31 (1975), 1-15.

[72] D. RESCH, Some Baeckland Transformations of Partial Differential Equations of Second Order, Syracuse University Thesis, Syracuse, 1950.

[73] L. BERS, Mathematical Aspects of Subsonic and Transonic Gasdynamics, Wiley, New York, N.Y., 1958.

[74] C. ROGERS AND J.G. KINGSTON, Non-dissipative magnetohydrodynamic flows with magnetic and velocity field lines orthogonal geodesics on a normal congruence, SIAM J. Appl. Math. 26 (1974), 183-195.

[75] H. GRAD, Reducible problems in magneto-fluid dynamic steady flows, Rev. Mod. Phys 32 (1950), 830-846.

[76] K. STEWARTSON, The dispersion of the current on the surface of a highly conducting fluid, Proc. Cambridge Philos. Soc. 53 (1957), 774-775.

[77] I.M. IŬR'EV, On a solution to the equations of magneto-gasdynamics, J. Appl. Math. Mech. 24 (1960), 233-237.

[78] C. ROGERS, H.M. CEKIRGE, AND A. ASKAR, Electromagnetic wave propagation in non-linear dielectric media, submitted for publication.

[79] D.L. CLEMENTS AND C. ROGERS, On anti-plane contact problems involving an inhomogeneous half-space, submitted for publication.

[80] S. BERGMAN, Integral Operators in the Theory of Linear Partial Differential Equations, Springer-Verlag, New York, N.Y., 1968.

SOME OLD AND NEW TECHNIQUES FOR THE PRACTICAL USE OF EXTERIOR DIFFERENTIAL FORMS*

Frank B. Estabrook

Jet Propulsion Laboratory
California Institute of Technology
Pasadena, California 91103

Such the use of forms
Peculiar in the realms of Space or Time:
Such is the throne which Man for truth amid
The paths of mutability hath built
Secure, unshaken, still; and whence he views
In matter mouldering, structures, the pure
    forms...

--J. M. W. Turner

I. INTRODUCTION

Cartan's calculus of exterior differential forms [1] is a convenient mathematical tool for systematically applying many otherwise disparate techniques of differential analysis. In the notational guise of modern differential geometry and mapping theory, this calculus has been used for deriving important global results for systems of ordinary differential equations; we believe it is at least as useful for expressing local, differential relations, and especially for treating systems of partial differential equations [2], [3], [4].

This is a brief summary of the differential form techniques we have developed and applied thus far. Using a convenient notation, we first summarize the algebra and calculus of forms. Unfortunately, there are many variations of notation and even of the words that go with the underlying concepts; we will try to indicate some of these as we go along. We always assume everything suitably compact, continuous, and differentiable, although it will be obvious that in

*This paper presents the results of one phase of research carried out at the Jet Propulsion Laboratory, California Institute of Technology, under Contract No. NAS7-100, sponsored by the National Aeronautics and Space Administration.

many cases differential form expressions are immediately convertible to integral results by the Stokes theorems, and then topics like allowable topologies and jump discontinuities could be discussed.

II. FORMS

An n-dimensional differentiable manifold can be most simply introduced as a Euclidean space with Cartesian coordinates x^i, $i = 1,\ldots,n$. This structure ensures satisfaction of all the local topological requirements for continuous point sets, so that we may proceed with the usual analytical and differentiation operations when dealing with sufficiently continuous functions of the x^i. But once the local point set topology has been defined, the particular variables x^i are not in any way taken as intrinsically preferred over any other set related to them by smooth non-singular transformation, and no metric, or measure of distance or interval, is necessarily given. In practice, the identification and postulation of sufficiently smooth independent coordinates to define a differentiable manifold is always done quite automatically in setting up a problem, when there is a clear physical model to consider.

So, let us consider an n-dimensional differentiable manifold spanned by a set of scalar fields x^i, $i = 1,\ldots,n$, each with a continuous range of values. A function $\phi(x^i)$ of those scalars is then geometrically represented as the family of (n-1)-dimensional surfaces obtained by setting $\phi(x^i)$ = constant. The total differential or gradient of a function ϕ is the prototype 1-form: at each point x^i the expression

$$d\phi = \frac{\partial\phi}{\partial x^i}\, dx^i \equiv \phi_{,i}dx^i$$

(we will use throughout the summation convention for repeated indices) describes the orientation of the (n-1)-surface through the point, and the spacing[1] there of successive members of the family, relative to the coordinate grid and scales set

---

[1]Or better, <u>density</u> in the "quotient" space whose points are the (n-1)-surfaces.

by the x^i. By the same token, the total differential dx^1 describes the coordinate family of (n-1)-surfaces x^1 = constant, dx^2 describes the family x^2 = constant, etc. Thus the field $d\phi$ is at each point a linear superposition of components on the <u>basis forms</u> dx^i there. These define a linear vector space at each point x^i, known as the cotangent space. In traditional tensor notation, $d\phi$ is described by the array of its components: $\phi_{,i} = \frac{\partial\phi}{\partial x^1},\ldots,\frac{\partial\phi}{\partial x^n}$. Note carefully then that $d\phi$ is a first order covariant tensor field; one should not think of it as an infinitesimal. A general 1-form, say ω, is an arbitrary linear superposition on basis 1-forms at each point: $\omega = f_i(x^j)dx^i$. One may think of a local, oriented, spaced set of surfaces (a "sandwich" structure) at each point of the x^i space; if these are to mesh from point to point to form a global family of (n-1)-surfaces, there must exist a function ϕ such that $\omega = d\phi$, i.e. we would have $f_i = \phi_{,i}$.

A (two parameter) family of (n-2)-surfaces could be described by setting $\phi(x^i)$ = constant, $\psi(x^i)$ = constant, where ϕ and ψ are functions leading to linearly independent 1-forms $d\phi$ and $d\psi$ at each point x^i. However, arbitrary functions $\phi'(\phi,\psi),\psi'(\phi,\psi)$ would do as well for describing the orientation if they lead to independent 1-forms that are linear combinations of $d\phi$ and $d\psi$. The essential algebraic content is expressed by the "exterior" product, written $d\phi\wedge d\psi$, where $\wedge$ denotes the completely antisymmetrized tensor product. This is a geometric object which describes at each point x^i the orientation, and spacing, of a family of (n-2)-surfaces. In tensor calculus this "2-form" $d\phi\wedge d\psi$ is described by its coordinate components, $\phi_{,i}\psi_{,j} - \phi_{,j}\psi_{,i}$. It is a completely antisymmetric second order covariant tensor, say τ_{ij} (of an algebraically special kind called "simple" or "monomial", since it satisfies $\tau_{ij}\tau_{k\ell}\varepsilon^{ijk\ell} = 0$). A general 2-form is an arbitrarily linear superposition on simple basis 2-forms: $f_{ij}(x^k)dx^i\wedge dx^j$.

General p-forms, $p = 1,\ldots,n$, are similarly defined with components on simple basis elements of lower dimensional surfaces, such as (for $p = 3$) $f_{ijk}dx^i\wedge dx^j\wedge dx^k$, etc. This last expression is unambiguous because the operation of exterior product, $\wedge$, is associative. All p-forms, $p = 1,\ldots,n$, at a point constitute an associative algebra--the Grassmann algebra--on the cotangent space.

The algebra of forms is constructed from scalar fields, and the "Eulerian" operation d that operates on them to generate 1-forms. It is shown in all standard references how this differential operation d is further applicable to forms of any order. In tensor or coordinate language, it generates the "curl"--a completely antisymmetrized set of partial derivatives of the components of a given form, readily shown to be the components of a form of the next higher order. An important theorem is that applying d twice one gets zero identically. A form is called "exact" if its exterior derivative is zero. The "natural" basis forms dx^i belonging to a coordinate frame x^i are obviously exact.

We have to remember thus far the rules for manipulation:

$$\omega\wedge\sigma = (-1)^{pq}\sigma\wedge\omega \quad ,$$

$$d(\omega\wedge\sigma) = d\omega\wedge\sigma + (-1)^{p}\omega\wedge d\sigma \quad ,$$

$$dd\omega = 0 \ ,$$

$$dc = 0 \ ,$$

where p and q are the orders of ω and σ, respectively, and c is a constant.

Although in tensor calculus 1-forms are called "covariant vectors", the terminology should perhaps be avoided because (contravariant) vectors are very different geometrical objects. (In Riemannian geometry one always has a way of associating 1-forms and vectors, using a metric tensor g_{ij} and its inverse g^{ij}.) On the other hand, the true vectors, which we consider next, do come close to the concept of an (n-1)-form. The latter describes, at a point, the orientation and spacing of a family--or congruence--of 1-dimensional submanifolds. By "spacing" is meant the density of the points representing the submanifolds in the (n-1)-dimensional "quotient space" of the congruence. If a density field in n-space were also given,[2] the (n-1)-form, divided by this, would describe the orientation of a congruence together with a magnitude which now is a measure of 1-dimensional extension or length at each point, and <u>this</u>, up to a simple choice of sense or direction, <u>is</u> the geometric description of a vector field.

III. VECTORS

A family of 1-dimensional submanifolds, "calibrated" with an internal coordinate laid out along their lengths, is a vector field $\vec{V}$. In any coordinate system x^i one has (in an older notation where by $\frac{d}{ds}$ one means "total" differentiation) a set of autonomous first-order ordinary differential equations

$$\frac{dx^i}{ds} = V^i(x^j) \quad ,$$

and the general integral of these is a parametric description of the submanifolds: $x^i = H^i(s,a_1,a_2,\ldots,a_{n-1})$. The dependence on the scalar field s fixes the internal calibration. In a notation to be introduced shortly, we would have $\vec{V}\cdot ds = 1$. $a_1,\ldots,a_{n-1}$ are $n-1$ (the maximum number) independent "constants of the motion", or "first integrals" or "co-moving coordinates"; $\vec{V}\cdot da_i = 0$. Any $n-1$ independent functions of the a_i would do as well. Moreover, s could be "set" differently on each line of the congruence, without altering the V^i, by writing $x^i = H^i(s' - p(a_1,\ldots,a_{n-1}),a_1,\ldots,a_{n-1}) = \bar{H}^i(s',a_1,\ldots,a_{n-1})$. Thus the functional form of the H^i is far from unique. We will see how a unique or canonical functional description of the congruence of 1-dimensional submanifolds can be set up using n constants of the motion, say $\bar{x}^i$, chosen so that $x^i = \bar{x}^i$ at $s = 0$. s is then treated as an external parameter, each fixed value of which corresponds to a point translation or mapping of the entire space along the congruence: $x^i \to \bar{x}^i$.

The functions V^i in the above set of equations are denoted "components" of $\vec{V}$ with respect to the x^i. In the sense that this is a set of directional

---

[2] For example, by the single non-vanishing coordinate component of an n-form, or by just the single quantity $\sqrt{|\det g_{ij}|}$ calculated from a metric tensor field.

derivatives of the x^i, along $\vec{V}$, these equations are best written as a set of Lie derivatives--we will return to that interpretation of a vector field shortly. There is, however, a local sense in which $\vec{V}$ gives an entity in an n-dimensional linear vector space attached to each point of the differentiable manifold. This is the <u>tangent space</u>, the objects (vectors) in which turn out to be algebraically dual to the objects (1-forms) in the cotangent space there. That is to say, we will find an operation of <u>inner product</u> or <u>contraction</u> between them: at a point, any vector $\vec{V}$ maps all 1-forms onto the real numbers (or conversely).

All vector fields $\vec{V}$ which have the same components V^i at a point $x_0^i = x^i(s_0,\dots)$ are, as an <u>equivalence class</u>, denoted (without changing the symbolism) just the vector $\vec{V}$ there. We can arrange that $s_0 = 0$ there for all of them. They are all mutually tangent, and in particular tangent to a family of 1-dimensional manifolds that is linear--straight--with respect to the x^i coordinates, so satisfying

$$x^i = x_0^i + \left.\frac{dx^i}{ds}\right|_{s=0} s = x_0^i + V^i(x_0^j)s \ .$$

Clearly, any such local vectors at x_0^i can be linearly superimposed, as with any fixed conventional value of s they are now seen as displacements in an affine space--again the tangent space. This construction provides the usual geometric image one has of vectors as superimposable displacements to neighboring points, but in the differentiable manifold itself this is only strictly true for infinitesimal values of s (unless one restricts oneself to a particular coordinate frame and those linearly related to it--again the affine geometry!).

Any coordinate frame x^i, which furnishes basis forms dx^i for the cotangent space, also furnishes dual basis vectors for the tangent space. We denote by $\vec{\lambda}_1$ the 1-dimensional submanifolds on which $x^2, x^3, \dots, x^n$ are constant, and on which we adopt a calibration so that $x^1 = s$. The rest of the $\vec{\lambda}_i$ are similarly defined. Any vector $\vec{V}$ then is expressed in the basis belonging to the coordinate system as $V^i\vec{\lambda}_i$, where the V^i are as before the components with respect to that same coordinate system.

At a point, say x_0^i, dx^1 describes the orientation of an (n-1)-surface through the point, on which $x^1 = x_0^1$ is constant, and describes as well the spacing of adjacent surfaces of the same family. Laying off an infinitesimal displacement $\epsilon\vec{\lambda}_i$ from the point we stay in the surface if $i \neq 1$, and we are brought to an adjacent surface if $i = 1$. The calibration of $\vec{\lambda}_1$ ensures that in the last case the value of x^1 on the adjacent surface is just $x_0^1 + \epsilon$. The <u>change in value</u> of x^1 along any vector is a scalar denoted the inner product, or contraction, of the vector and dx^1; it is denoted by an interposed $\cdot$ or $\lrcorner$. For $\epsilon\vec{\lambda}_i$ we have just found $\epsilon\vec{\lambda}_i \cdot dx^1 = \epsilon\delta_i^1$; in general we have

$$\vec{\lambda}_i \cdot dx^j = \delta_i^j \ .$$

We then find $V^i = \vec{V} \cdot dx^i$ and $(V^i\vec{\lambda}_i) \cdot (f_j dx^j) = V^i f_i$. Both forms and vectors enter the contraction operation linearly.

One justifies, and remembers, the following general rules for contraction of a vector with higher rank forms:

$$(\vec{V} + \vec{W}) \cdot \omega = \vec{V} \cdot \omega + \vec{W} \cdot \omega \ ,$$

$$(f\vec{V}) \cdot \omega = f\vec{V} \cdot \omega \ ,$$

$$\vec{V} \cdot (\omega \wedge \sigma) = (\vec{V} \cdot \omega) \wedge \sigma + (-1)^p \omega \wedge (\vec{V} \cdot \sigma) \ ,$$

where p is the rank of ω.

In modern texts vectors are defined in a different fashion. A vector field is the generator of a 1-parameter continuous group of transformations of the space, a diffeomorphism mapping it onto itself. More explicitly, the functions $V^i(x^j)$ can be regarded as generating a 1-parameter family of scalar fields $\bar{x}^i(x^j,s)$, of which $\bar{x}^i(x^j,0) = x^i$, $\left.\frac{\partial \bar{x}^i}{\partial s}\right|_{s=0} = V^i$, and the higher-order Taylor series coefficients are determined from these by the group property. A symbolic expression of this transformation group is

$$\bar{x}^i = F^i(x^j,s) = \exp(s\mathcal{L}_{\vec{V}})x^i \ ,$$

which implies the transitive rule

$$F^i(F^j(x^k,s),\bar{s}) = F^i(x^j,s+\bar{s}) .$$

The inversion is $x^i = F^i(\bar{x}^j,-s)$. In x^i space the $\bar{x}^i$ are n first integrals of $V^i(x^j)$, and this relation is a canonical functional description of the congruence of calibrated lines. By imposing one arbitrary functional relation between the $\bar{x}^i$, for each value of the parameter s we would find an (n-1)-manifold in the space of the x^i, and so s would then again appear as a scalar field there in an internal parametric description.

$\underset{\vec{V}}{£}$ is a linear differential operator which, when applied to x^i, gives the set of scalars V^i; this follows from the above, or simply from calculating

$$\underset{\vec{V}}{£}\, x^i \equiv \lim_{s\to 0} \frac{\bar{x}^i - x^i}{s} = V^i .$$

We ignore all global and topological niceties of the modern understanding of diffeomorphisms of differentiable manifolds, and simply assert that (in most neighborhoods) the linear operators $\underset{\vec{V}}{£}$ on a differential manifold are abstractly equivalent to the vector fields $\vec{V}$ as we previously introduced them; they have the same coordinate components, the scalar fields $\underset{\vec{V}}{£}\, x^i = V^i = \vec{V}\cdot dx^i$. Since any scalar ϕ could be a "coordinate", we have in fact

$$\underset{\vec{V}}{£}\,\phi = \vec{V}\cdot d\phi \quad .$$

A continuous group of mappings of a differentiable manifold onto itself maps at the same time all geometric objects in it. Thus $\underset{\vec{V}}{£}$ can operate on scalars, forms, and vectors. The result is in each case necessarily of the same tensor character, and is called the <u>Lie derivative</u> with respect to $\vec{V}$. Applied to scalars, it has been seen to be the usual substantial or material derivative of fluid mechanics. (We previously described 1-forms as "Eulerian"; we can now make the dual remark that vectors are "Lagrangian.") The directional or Lie derivative of forms, or other tensors, is less familiar than it should be, as it provides a very useful way of understanding, and remembering, the otherwise somewhat mysterious "correction terms due to the change of reference frame" that arise in all Lagrangian formulations for tensor fields. The key property of $\underset{\vec{V}}{\pounds}$ is that it is a derivation operator obeying Leibniz' rule. While we have not shown all this in detail here, the needed formulae for practical manipulations can now be quickly obtained.

First, for exact forms dx^i, since $d\bar{x}^i = dx^i + s dV^i + \ldots$,

$$\underset{\vec{V}}{\pounds} dx^i = \lim_{s\to 0} \frac{d\bar{x}^i - dx^i}{s} = dV^i = d\underset{\vec{V}}{\pounds} x^i .$$

For a general 1-form $\omega = f_i dx^i$ we then have

$$\begin{aligned}\underset{\vec{V}}{\pounds}\omega &= (\underset{\vec{V}}{\pounds} f_i) dx^i + f_i \underset{\vec{V}}{\pounds} dx^i \\ &= (\vec{V}\cdot df_i) dx^i + f_i dV^i \\ &= \vec{V}\cdot d\omega + (\vec{V}\cdot dx^i) f_i + d(f_i V^i) - V^i df_i \\ &= \vec{V}\cdot d\omega + d(\vec{V}\cdot\omega) ,\end{aligned}$$

and in fact this last relation holds for any form ω, of any higher rank. If ω is exact, say $d\alpha$, it says that $\underset{\vec{V}}{\pounds}$ and d are always commuting operations:

$$\underset{\vec{V}}{\pounds} d\alpha = d(\vec{V}\cdot d\alpha) = d(\vec{V}\cdot d\alpha + d\{\vec{V}\cdot\alpha\}) = d\underset{\vec{V}}{\pounds}\alpha .$$

The Lie derivative of one vector field $\vec{W}$ with respect to another $\vec{V}$ can be found most readily in terms of components:

$$\vec{V}\cdot dW^i = \underset{\vec{V}}{\pounds} W^i = \underset{\vec{V}}{\pounds}(\vec{W}\cdot dx^i) = (\underset{\vec{V}}{\pounds}\vec{W})\cdot dx^i + \vec{W}\cdot\underset{\vec{V}}{\pounds} dx^i = (\underset{\vec{V}}{\pounds}\vec{W})^i + \vec{W}\cdot d\vec{V}^i$$

or

$$(\underset{\vec{V}}{\pounds}\vec{W})^i = \vec{V}\cdot dW^i - \vec{W}\cdot dV^i .$$

This is antisymmetric in the two vectors. It is customary to denote it as the Lie product, or bracket or commutator, $[\vec{V},\vec{W}]$, which is quite appropriate since we here regard the vector fields as linear operators. This also provides a language for geometrically expressing the properties of Lie groups and algebras.

As for "natural" basis vectors $\vec{\lambda}_i$ (i.e. derived from coordinates), they satisfy $[\vec{\lambda}_i,\vec{\lambda}_j] = 0$. A common "operator" notation for the basis vectors is $\vec{\lambda}_i = \partial/\partial x^i$.

One finally needs to remember only four results: that $\underset{\vec{V}}{£}$ is a derivation both for forms

$$\underset{\vec{V}}{£}(\omega\wedge\sigma) = (\underset{\vec{V}}{£}\omega)\wedge\sigma + \omega\wedge\underset{\vec{V}}{£}\sigma$$

and for contractions

$$\underset{\vec{V}}{£}(\vec{W}\cdot\omega) = [\vec{V},\vec{W}]\cdot\omega + \vec{W}\cdot\underset{\vec{V}}{£}\omega \ ,$$

that the Jacobi identity holds for vectors,

$$[\vec{U},[\vec{V},\vec{W}]] + [\vec{V},[\vec{W},\vec{U}]] + [\vec{W},[\vec{U},\vec{V}]] = 0 \ ,$$

and that, operating on any geometric object,

$$\underset{\vec{U}\vec{V}}{££} - \underset{\vec{V}\vec{U}}{££} = \underset{[\vec{U},\vec{V}]}{£} \ .$$

IV. INTEGRATION

A p-form ϕ can be integrated over a p-dimensional subspace V. A commuting set of p independent vector fields lying in V (hence said to be p-forming), say $\vec{V}_1,\ldots,\vec{V}_p$, can be used to divide up the subspace into infinitesimal p-parallelopipeds with edges $\epsilon_1\vec{V}_1,\ldots,\epsilon_p\vec{V}_p$,[3] and for each of these the form ϕ can be evaluated by contraction: $\epsilon_1\ldots\epsilon_p\vec{V}_1\cdot(\vec{V}_2\cdot(\ldots(\vec{V}_p\cdot\phi)\ldots))$. An integral is then defined by summing all these to the (p-1)-dimensional bounding manifold (denoted ∂V), and suitably going to a limit. The symbol for this is just $\int_V\phi$ as (due to the antisymmetry of the p-form) the result is independent of the auxiliary mesh.

The generalized Stokes theorem can be proved by laboriously considering

[3]Thus we are assuming for convenience and brevity that V is "oriented".

the integration just described. One need only remember that if the topology is simple (orientable)

$$\int_V d\sigma = \int_{\partial V} \sigma ,$$

where σ is a (p-1)-form, and $d\sigma$ is an exact p-form. If σ is exact, in V, so $d\sigma = 0$, we have $\int_{\partial V} \sigma = 0$: the integral of an exact form over a closed manifold is zero.

It may be shown--indeed, is an alternate way of defining Lie derivatives of forms [4]--that if the points of the integration boundary ∂V are infinitesimally displaced by $\epsilon\vec{v}$ the change in $\int_V \phi$ is $\epsilon \int_V \pounds_{\vec{v}} \phi$. This is a very useful rule for variational calculations.

V. THE SOLUTIONS OF PARTIAL DIFFERENTIAL EQUATIONS AND THE INTEGRAL MANIFOLDS OF IDEALS OF FORMS

An immersed manifold or subspace V_p of p-dimensions can be described by giving $n - p$ variables as functions of p others: this is the geometrical distinction between dependent and independent variables. We will in this section define the so-called integral manifolds V_p belonging to--defined by--a given set of exterior differential forms. These are solutions of resulting sets of first-order partial differential equations, and it follows that the many systematic techniques for manipulating differential forms consequently have much practical use in the analysis of sets of partial differential equations. In the next section we will consider the construction of the so-called "regular" integral manifolds, which is Cartan's generalization of the Cauchy-Kowalewski theorem to sets of first-order partial differential equations. We will come upon criteria for a set of forms to be "well set", viz. that its regular integral manifolds include almost all of the usually required solutions of a given set of partial differential equations. (The exceptions are so-called "singular" integral manifolds describing certain characteristic or exceptional solutions; on the other hand, the regular integral manifolds can include generalized solutions introduced by Lie in which the roles of some independent and dependent variables are interchanged.) Such considerations may eventually lead to sharper criteria than we can at present formulate for the proper choice of sets of forms to represent partial differential equations.

Consider as an example that we are given, in n-space, a set of $m < n$ independent 1-forms ω^A, $A = 1,\dots,m$, together with their exterior derivatives, the 2-forms $d\omega^A$. (For simplicity we do not explicitly consider any other 2-forms, independent 3-forms or higher, to be given, but it should be clear, how readily they could be included.) Denote this a Pfaffian system. If we find a set of independent vector fields $\vec{V}_1,\dots,\vec{V}_p$, all of which lie in a family of p-dimensional subspaces V_p (this requires $[\vec{V}_i,\vec{V}_j] = f^k_{ij}\vec{V}_k$ and is denoted p-forming[4]), and which are such that they each <u>annul</u> (give zero when contracted on) any ω^A

$$\vec{V}_i\cdot\omega^A = 0,$$

then by operating with $\pounds_{\vec{V}_j}$ it is quickly found that also

$$\vec{V}_i\cdot\vec{V}_j\cdot d\omega^A = 0 .$$

Any vector in V_p annuls any 1-form in the set, and taken two by two, any pair of vectors annul the 2-forms $d\omega^A$. Such a V_p--containing $\vec{V}_1,\dots,\vec{V}_p$--is called an integral manifold, or integral subspace, of the set ω^A, $d\omega^A$. Our definition is easily extended to sets including forms of higher rank; again their exterior derivatives can be included. Any set is denoted "closed" if it includes the exterior derivatives of all forms in it, and we will henceforth only consider closed sets. In general they will have integral manifolds of various dimensionalities $p = 1,2,\dots$.

In modern texts a subspace V_p is described as a map ϕ taking points of the p-manifold V_p to the immersing n-manifold $(n \geq p)$. Forms--

---

[4]By rescaling the $\vec{V}_j$ the f^k_{ij} could be made zero--the $\vec{V}_j$ would then form a mesh of p-parallelopipeds like the $\vec{\lambda}_j$ of a coordinate frame in V_p.

themselves locally described as arrays of components on subspaces--are readily shown to transform only under the inverse map ϕ^*, i.e. forms ω^A and $d\omega^A$ in the n-manifold may be "pulled back" (or mapped, or induced, or sectioned) by ϕ^* onto V_p, where they may be denoted $\tilde{\omega}^A$ and $d\tilde{\omega}^A$.[5] They are there expressible solely in terms of a p-dimensional basis of 1-forms. In the present case, since the results of contracting ω^A with any vector in V_p and $d\omega^A$ with any pair of vectors in V_p are both zero, the condition determining integral submanifolds can be restated simply as $\tilde{\omega}^A = 0$ and $d\tilde{\omega}^A = 0$. So finally, for a general set of forms, of no matter what rank, an integral manifold is defined by requiring them all to be zero when pulled back into it. Since d and the inverse mapping operation ϕ^* commute, the sets of forms considered may always be taken closed.

Since the forms ω^A, $d\omega^A$, etc. only enter the definition of integrable manifolds through linear homogeneous equations, it is clear that any algebraically equivalent set of forms may be used. It should, in fact, perhaps have been stated at the outset that it is an entire <u>ideal</u> of the cotangent space Grassmann algebra, generated by the given forms, here ω^A and $d\omega^A$, that is annulled--pulled back to vanish in any integral manifold V. That is to say, all 1-forms such as $f_A\omega^A$, 2-forms $\psi_A \wedge \omega^A + f_A d\omega^A$, etc. ($f_A$ being arbitrary functions, ψ_A being arbitrary 1-forms, etc.) will obviously be annulled, when ω^A and $d\omega^A$ are. For a Pfaffian system any set of $n - m$ linearly independent 1-forms algebraically formed from the ω^A, together with its closure 2-forms, generates the same ideal: $f^B_A\omega^A$, $d(f^B_A\omega^A)$, where the functions f^B_A are such that $|f^B_A| \neq 0$.

If we introduce p autonomous independent variables y^μ, $\mu = 1,\ldots,p$ as a basis in every V_p, we have, if $\omega^A = \omega^A_i(x^j)dx^i$, $d\omega^A = \omega^A_{i,j}dx^j \wedge dx^i$, that $\tilde{\omega}^A = \omega^A_i \frac{\partial x^i}{\partial y^\mu} dy^\mu$, $d\tilde{\omega}^A = \omega^A_{i,j} \frac{\partial x^j}{\partial y^\nu} \frac{\partial x^i}{\partial y^\mu} dy^\nu \wedge dy^\mu$. Setting these equal to zero, term by term, gives a coupled set of partial differential equations for n dependent

---

[5]For any p-form ϕ it follows that $\int_{V_p} \phi = \int_{V_p} \tilde{\phi}$.

variables x^i as functions of the y^μ. Any algebraically equivalent set of forms, pulled back to zero, will give an equivalent coupled set. If p of the x^i can still vary freely in the integral manifolds, they could have been adopted as the independent variables and non-autonomous equations would result for $n - p$ dependent variables. In either event, note that all integrability conditions are explicitly included in the resulting set of first-order partial differential equations.

VI. THE REGULAR INTEGRAL MANIFOLDS AND CARTAN'S CHARACTERS

We are often asked about the converse problem: given a set of first-order partial differential equations, how to choose a set of forms,to represent the desired solutions. Cartan's generalization of the Cauchy-Kowalewski theory gives criteria guiding this selection. The selection is even then far from unique; nevertheless we have found in practice that all such sets of forms are highly useful. We suspect that more succinct criteria will emerge after further experience.

Cartan considers a sequence of families of successively higher-dimensioned integral manifolds denoted, from their sequential construction from boundary conditions, as "regular". Algebraic criteria emerge for the "degrees of freedom" in the construction of the regular integral manifolds, in terms of so-called "characters", obtained from the ranks of certain sets of linear algebraic equations. In the maximum dimensional family (which has the maximum number of independent variables), those variables which can still be freely varied are denoted as being "in involution" (this terminology is not to be confused with that of Hamiltonian theory). We consider a set of forms to be "well set" for describing a given set of first-order partial differential equations, when the maximal regular integral manifolds of the set of forms are the desired solutions of the equations, and the given independent variables are in involution. These criteria can be determined by purely local arithmetic calculations with the characters.

Begin, at some general point having coordinates x^i, with a vector $\vec{V}_1$ chosen such that the m conditions

$$\vec{V}_1 \cdot \omega^A = 0$$

are satisfied. These are a set of homogeneous linear algebraic equations relating the n components $V_1^i = \vec{V}_1 \cdot dx^i$ of $\vec{V}_1$. To conform with Cartan's notation, denote the rank of this first algebraic set that we must consider by s_0 (=m), so $\ell_1 = n - s_0 > 0$ components of $\vec{V}_1$ may be chosen arbitrarily. (We take this rank to be at its maximum at x^i, and in a neighborhood of x^i. Thus we ignore all special cases that might arise in this and the following, for clarity of exposition of the central ideas.)

We can now advance to the point $\bar{x}^i = x^i + \epsilon V_1^i$ where ϵ is infinitesimal, and again make ℓ_1 choices (which can differ from those first made to order ϵ) to determine a vector $\vec{V}_1$ at $\bar{x}^i$. This can be done in repeated fashion, varying the V_1^i each time to the same order ϵ, so by a limiting process based on the sequence of such step-by-step choices we construct a continuously imbedded 1-dimensional integral manifold $V_1 = \{x^i : x^i = f^i(s), df^i/ds = V_1^i(s)\}$. In constructing such a V_1 we have made ℓ_1 choices of functions of one variable while integrating a set of ordinary differential equations from given initial conditions.

At each point of a V_1 we can choose a vector $\vec{V}_2$ such that

$$\vec{V}_2 \cdot \omega^A = 0 \ ,$$

$$\vec{V}_2 \cdot \vec{V}_1 \cdot d\omega^A = 0 \ .$$

(If there were other 2-forms given, in addition to the $d\omega^A$, we would include them also in the last equation.) The rank of this set of homogeneous linear equations for the n components of $\vec{V}_2$ is not less than s_0; we denote it, with Cartan, $s_0 + s_1$. So we can in general choose $\ell_2 = n - (s_0+s_1)$ members of the set V_2^i arbitrarily. We remark, however, that the set V_1^i is of course a solution of the above equations, and so is included in the ℓ_2 linearly independent possible choices. If it were the only possible choice, we would not be able to proceed with the construction of a V_2, so let us assume $\ell_2 > 1$.

We can now, from each point x^i of V_1, advance to a neighboring point $x^i + \zeta V_2^i$ not in V_1. ζ is again infinitesimal. A new, neighboring V_1 is thus constructed point-wise; it will have a natural calibration carried along from the ϵ steps on the first V_1, i.e. a new $\vec{V}_1$ is given at each point. We will show in a moment that this new V_1 is again an integral manifold, i.e. that still $\vec{V}_1 \cdot \omega^A = 0$. A varied choice of $\vec{V}_2$, within ℓ_2 degrees of freedom, can be made on the new V_1, and indeed on each new V_1 as the construction now steps along in ζ intervals. Again by a limiting argument we assert that a V_2 is thus constructed: $V_2 = \{x^i : x^i = f^i(s,t), \frac{\partial f^i}{\partial s} = v_1^i(s,t), \frac{\partial f^i}{\partial t} = v_2^i(s,t)\}$. We have now made ℓ_2 further choices of functions, this time functions of two variables.

$\vec{V}_1$ has not been free in this last construction, rather it has been defined at each step so as to be "carried along" $\vec{V}_2$, forming infinitesimal parallelograms that close. That is, we have determined $\vec{V}_1$ throughout V_2 by the condition $[\vec{V}_1, \vec{V}_2] = 0$. This is also clearly the integrability requirement on $f^i(s,t)$. Its imposition is the key to Cartan's construction. For consider the resulting scalar fields $\vec{V}_1 \cdot \omega^A$ in V_2. The Lie derivatives of these along $\vec{V}_2$ are $\pounds_{\vec{V}_2} (\vec{V}_1 \cdot \omega^A) = [\vec{V}_2, \vec{V}_1] \cdot \omega^A + \vec{V}_1 \cdot \{d(\vec{V}_2 \cdot \omega^A) + \vec{V}_2 \cdot d\omega^A\}$; but these are zero, term by term. Thus since $\vec{V}_1 \cdot \omega^A = 0$ on the "boundary"--the initial V_1--they remain zero throughout V_2. Thus we have indeed constructed V_2 so that any vectors in it, contracted on the ω^A, give zero, and so that any pair contracted on the $d\omega^A$ gives zero. It is an integral manifold.

The construction continues. Given a V_2, choose at each point a $\vec{V}_3$ such that

$$\vec{V}_3 \cdot d\omega^A = 0 ,$$

$$\vec{V}_3 \cdot \vec{V}_1 \cdot d\omega^A = 0 ,$$

$$\vec{V}_3 \cdot \vec{V}_2 \cdot d\omega^A = 0 .$$

(If there were independent 3-forms in the given closed ideal of forms, obviously the required additional linear homogeneous equations for V_3^i would be included.) In calculating the rank of this linear homogeneous set for the V_3^i one must take account of all information known about the coefficients, e.g. that $\vec{V}_1 \cdot \vec{V}_2 \cdot d\omega^A = 0$, etc. The algebraic rank is denoted $s_0 + s_1 + s_2$; $\ell_3 = n - (s_0 + s_1 + s_2)$ independent solutions can be arbitrarily chosen. If $\ell_3 > 2$, a V_3 can be constructed by making ℓ_3 choices of functions of 3 variables; $\vec{V}_1$ and $\vec{V}_2$ are constructed at the same time throughout the V_3 such that $[\vec{V}_1,\vec{V}_3] = 0$, $[\vec{V}_2,\vec{V}_3] = 0$, $[\vec{V}_1,\vec{V}_2] = 0$, and V_3 is shown to be an integral manifold. The proof at each stage depends essentially on the fact that the ideal is closed.

The positive integers $s_0, s_1, s_2, \ldots$ are denoted the <u>characters</u>. They are numerical invariants of the ideal of forms. Since we have $\ell_p \leq \ell_{p-1}$, $p = 1,2,\ldots,$ and moreover require at each step $\ell_p > p - 1$, the sequential construction must terminate: there must be a maximum dimension of the families of regular integral manifolds. Let the largest value of p be denoted g, the <u>genus</u>; then $\ell_g = n - (s_0+s_1+\ldots+s_{g-1}) > g - 1$, and $\ell_{g+1} = n - (s_0+s_1+\ldots+s_{g-1}+s_g) \leq g$, $\ell_{g+1} \leq \ell_g$. If $\ell_g > g$, the integration scheme remains under-determined at the last step: the equivalent set of partial differential equations for $n - g$ dependent variables and g independent variables will be under-determined in that arbitrary functions of the independent variables will enter the solution.

The case of most interest is when $\ell_g = g$: there is then no freedom in the last construction--V_g is determined by properly set data on its boundary. In this case the above inequalities require the precise relation

$$n - g = s_0 + s_1 + \ldots + s_{g-1} .$$

This is an important diagnostic test for a closed ideal of forms to meet, to be chosen to represent a proper set of partial differential equations: the sum of the Cartan characters $s_0 + s_1 + \ldots + s_{g-1}$, where g is the number of independent variables, must equal the number of dependent variables, $n - g$. Conversely, it seems to us a criterion for a "proper" set of partial differential equations that such an ideal can be found to represent it.

VII. INTEGRABLE SYSTEMS

If affine transformations of the 1-forms of a closed ideal can be found which yield exact 1-forms (i.e. if there are exact 1-forms in the ideal) the ideal is partially integrable. This case is concerned with the existence of enveloping manifolds (which arise for Pfaffian systems when $s_1 < s_0$). They are found by considering any $n - m$ independent vectors $\vec{\lambda}_\pi$, $\pi = 1,\ldots,n-m$, which annul the given ω^A: Consider any Lie product $[\vec{\lambda}_\pi,\vec{\lambda}_\sigma]$; if it is independent of the $\vec{\lambda}_\pi$, define it again to be a $\vec{\lambda}_\tau$, say. It will not annul the ω^A, but no matter, continue taking all possible Lie products of $\vec{\lambda}_\tau$'s until closure is achieved. Denote the dimensionality of that family of manifolds to be $n - q \geq n - m$. If $q > 0$, we have envelopment: the manifolds can be described by setting:

$$\phi^1 = \text{constant},$$
$$\vdots$$
$$\phi^q = \text{constant},$$

and the exact (!) 1-forms $d\phi^1,\ldots,d\phi^q$ are dual to all the $\vec{\lambda}_\tau$, and in particular to the original $\vec{\lambda}_\pi$:

$$\vec{\lambda}_\pi \cdot d\phi^b = 0, \; b = 1,\ldots,q, \quad \pi = 1,\ldots,n-m .$$

But this is now saying that the exact forms $d\phi^b$ must be in the ideal of the ω^A:

$$d\phi^b = f^b_A\omega^A \ , \quad b = 1,\ldots,q, \ q \leq m \ .$$

The f^b_A are integrating factors, and by putting the integrals ϕ^b = constant, our original set of forms can be reduced to a set in $n - q$ dimensions; the only problem is to ascertain that we have found the maximum value of q, or the minimum dimension of the enveloping manifolds.

An alternate approach to finding exact 1-forms in an ideal is just to consider the equations that result from expanding

$$df_A \wedge \omega^A + f_A d\omega^A = 0 \ .$$

These are an overdetermined set of linear partial differential equations for functions $f_A(x^i)$. If solutions exist, for each independent solution one can find a ϕ (up to arbitrary constant). $d\phi$ can be used, instead of one of the ω_A it depends on, as a basis for the ideal. The search for exact forms in a given ideal can be very useful also for higher rank forms. We will discuss this when we consider conservation laws.

For a Pfaffian system, in the extreme case when we find $q = m$, the number of independent 1-forms, we would find all the $d\omega^A$ in the ideal of just the ω^A--they contribute no independent equations to Cartan's constructions and could be dropped, $s_0 = m$, $s_1 = s_2 = \ldots = s_{n-m} = 0$. This is called a "completely integrable Pfaffian system." It could alternatively be described as $n - m$ vector fields--coupled ordinary differential equations without further integrability conditions.[6] The maximum regular integral manifolds then are of dimension $g = n - m$, and so are the minimum enveloping manifolds--they coincide! In this way we have come upon the Frobenius theorem.

VIII. CHARACTERISTIC VECTORS

In the remainder of these notes we assume a reasonably well-set closed ideal of forms to be presented for manipulation. We will discuss characteristic

6 We met such integrable systems in Cartan's construction of regular integral manifolds.

vectors, isovectors and the isogroup, linearity, similarity solutions, conservation laws, and prolongation structure of potentials and pseudopotentials. This is mainly dictated by our recent research experience, and is by no means exhaustive of possible systematic techniques with forms.

A characteristic vector $\vec{W}$ is defined by the property that contracting it on any form in a closed ideal of forms gives again a form in the ideal. The components of such a vector, together with a lot of undetermined functions that enter, satisfy linear algebraic equations. For example, if the ideal consists of 1-forms ω^A and 2-forms α^i, we have

$$\vec{W}\cdot\omega^A = 0 ,$$

$$\vec{W}\cdot\alpha^i = \psi^i_A\omega^A = 0 \text{ (mod ideal)} .$$

The ψ^i_A are undetermined functions. Given another characteristic vector, $\vec{U}$, we find, by operating on the above defining equations with $\pounds_{\vec{U}}$ that $[\vec{U},\vec{W}]$ is again a characteristic. Thus all the characteristics together are subspace-forming. The important theorem, easily seen by considering the Cartan construction, is that adjoining a characteristic direction to each point of a regular integral manifold gives again a regular integral manifold of, perhaps, one higher dimension. Thus the maximal regular integral manifolds must contain all the characteristics.

This last is the root of the variational principle of classical mechanics [5],[6]. For there we are given a "Hamiltonian structure" of forms

$$dH ,$$

$$dS - p_i dq^i ,$$

$$dp_i \wedge dq^i \quad i = 1,\ldots,n ,$$

where $H = H(p_i,q^i)$ is the Hamiltonian. We are in a $(2n+1)$-dimensional space, of p_i, q^i, S; $s_0 = 2$, $s_1 = s_2 = \ldots = s_{n-1} = 1$. The single characteristic vector field traces the classical trajectories, and the maximal regular integral manifolds, of dimension $g = n$, are the solutions of the Hamilton-Jacobi equation.

IX. ISOVECTORS

A generalization of characteristic vectors is the concept of isovector, $\vec{V}$--now it is the Lie derivative of any form ω^A or α^i with respect to $\vec{V}$ that is required to be in the ideal [7]:

$$\underset{\vec{V}}{\pounds}\omega^A = 0 \text{ (mod ideal)} ,$$

$$\underset{\vec{V}}{\pounds}\alpha^i = 0 \text{ (mod ideal)} .$$

(For a Pfaffian system the second of these is a consequence of the first.) The $\vec{V}$ are solutions of an overdetermined set of linear partial differential equations, and may be added with arbitrary constant factors. All $\vec{V}$ together generate the isogroup of the ideal.

It is with the isovectors that one arrives at the subject of this research workshop: contact transformations. (The representation of a higher-order partial differential equation by an ideal of forms requires that it first be converted to a set of first-order equations. We take it that the traditional adjective contact comes to little more than that.) Now a set of forms, along with all their families of integral manifolds, can be continuously transformed--varied--by mappings generated by vector fields in the space of all variables. We have seen that the Lie derivative is the proper description of this process. If the algebraic ideal of forms is kept invariant, the integral manifolds must all be simultaneously mapped into one another. We must have, in the generators of the resulting special group of mappings (the isovectors), the infinitesimal invariance transformations originally due to S. Lie. A number of examples of isovector calculations are given in [7].

Cartan seems to have missed these very useful auxiliary vectors because of his reluctance to use vectors at all--for example, he never defines Lie differentiation. Indeed Cartan even treats Lie groups exclusively with forms dual to the transformation generators used by Lie.

Auxiliary forms are discussed by Cartan for the process of prolongation: the simultaneous introduction of new forms and variables (which are essentially higher derivatives). By this, he showed that poorly-set ideals--ones where the integral manifolds of interest are not regular--can be extended into well-set ones. We have been concerned with the behavior of the isogroup under another sort of prolongation--the introduction of new forms and variables which are kinds of potentials--this will be discussed below. Under suitable prolongation the isogroup can become richer: in some sense the increased dimensionality of the space of variables, and the better-set prolonged ideal of forms in it, allows more freedom in associating neighboring integral manifolds into variational families of solutions. This is dramatically shown by setting up forms for the Burgers equation in the usual way, and finding their isogroup. It is finite, expressing scale invariance, and other less obvious invariances. After finding an exact 2-form in the ideal, and prolonging the original set with a corresponding potential, the isogroup becomes much larger--in fact, infinite. The infinite number of new isovectors are now themselves generated by solutions of a second-order linear equation--and correspondingly a superposition rule is found for solutions of the forms. The presence of the Cole-Hopf linearizing transformation is thus signaled by the behavior of the isogroup, and a search quickly discovers the proper dependent and independent variables to use in the enlarged space, in terms of which the linearity now present is made obvious.

X. SIMILARITY SOLUTIONS

If a general isovector is chosen, it will involve a number of arbitrary parameters equal to the dimensionality of the isogroup. If the isovector is not just a characteristic vector, a number of new forms can then be found by contraction on all forms in the ideal. Adding these to the ideal, a larger closed ideal of forms in the same space is found, the integral manifolds of which are a subset of those of the original ideal. These are the most general similarity solutions--and can be quite general when compared to those found intuitively with just one transformation invariance such as scaling. The integration to find the similarity solutions will involve one less independent variable, since the larger ideal now has an imposed characteristic vector.

XI. CONSERVATION LAWS [7]

If we can find an exact k-form, $1 < k \leq g$ in a well-set ideal of genus g (g independent variables in the maximal regular integral manifolds), we have found a conservation law for the set of partial differential equations. Denote the exact k-form by $d\psi$, where ψ is a (k-1)-form (determined up to the d of an arbitrary (k-2)-form), and from Stokes theorem we have

$$\int_V d\psi = \int_{\partial V} \psi$$

for any k-volume V, bounded by the closed (k-1)-manifold ∂V. If V lies in an integral manifold, $d\tilde{\psi}$ (pulled back into it) is zero. Thus we have $\int_{\partial V} \tilde{\psi} = 0$ over any closed (k-1)-manifold in an integral manifold. Seen in the g-space of independent variables, this is an integral conservation law, nontrivial if ψ itself is not in the ideal (for then $\tilde{\psi} = 0$).

XII. AUXILIARY FORMS FOR PROLONGATION

We have found conservation laws in the case $k = 2$ to allow a useful prolongation of the ideal [8]. When one can be found, we add the 1-form

$$dy + \psi$$

and at the same time admit the new variable y. Since the d of this form is already in the ideal, the prolonged ideal still has the same genus and independent variables. It can be said that $dy + \psi$ is a potential form, or a "conserved current."

It is of course quite possible that more than one such conservation law (or none) may be found--again, as in the finding of isovectors, one is integrating sets of overdetermined linear partial differential equations for auxiliary functions F_i defined from

$$\psi = F_i \, dx^i$$

and satisfying

$$d\psi = 0 \text{ (mod ideal)} .$$

The obvious generalization is to search for conservation laws--and further prolongations--of an already prolonged ideal. This sequential search, when successful, yields more conservation laws, the auxiliary functions of which depend on the previously introduced potentials $y_1, \ldots,$ (but not, at any step, on the latest y just to be introduced). The final generalization is to search for all of these at once--and indeed allow at any step all the y's to be involved in ψ. That is, we have finally come upon the very useful concept of pseudopotentials: a set of (an unspecified number of) 1-forms $dy_\alpha + F_{i\alpha} dx^i$ (where the $F_{i\alpha}$ are functions of the initial variables and all the y_α) whose exterior derivatives are in the new ideal (i.e. when prolonged with all the new 1-forms). With this generalization, we now obtain nonlinear equations--rather than linear ones--for the unknown functions $F_{i\alpha}$ in the ψ_α . We regard it as an exciting discovery that these can sometimes be thrown into the form of commutator equations between vector fields in the space of the y_α. Such a prolongation structure is then integrable by techniques familiar to group theorists, in particular the use of so-called linear or matrix representations.

XIII. THE KORTEWEG-DeVRIES EQUATION

In the case of the Korteweg-deVries (KdV) equation, formulated as a closed ideal of three 2-forms in a space of 5 dimensions, we have discussed much of the above in a recent paper [8]. We have since found the resulting prolongation structure to have the fascinating property that its own automorphism group is isomorphic to the simplest isogroup of the KdV equation [7]. It is a non-abelian 2-parameter group expressing scaling and Galilean invariances. Now similarity transformation of the prolongation structure can quickly be shown not to change the prolonged ideal of forms (it is equivalent to coordinate transformation in the space of the y's), but the automorphism transformation does--in general it will introduce two arbitrary parameters. The simplest non-trivial linear representation of the structure of the KdV that we have discovered so far is in terms of 2×2 matrices. This representation is quite degenerate--and it is also degenerate in that the automorphism group only introduces one arbitrary parameter into it. The result is a derivation of the two well-known coupled linear equations of the inverse scattering method, which indeed have one arbitrary parameter (the so-called eigenvalue λ). We are fascinated at the challenge of finding a 3×3 (or higher) representation which would not be degenerate under the automorphism group, and so which would contain two non-trivial parameters derived ultimately from the isogroup. The forms for the KdV equation would, for example, be a Pfaffian set of 3 1-forms and their 3 exterior derivatives, in an 8-dimensional space of u, $z = u_x$, $p = z_x$, x, t, y_1, y_2, y_3, the 1-forms linear and homogeneous in the y's and involving two arbitrary parameters. What would its isogroup be? And what form does the Bäcklund transformation for the KdV equation [9] take, for solutions of such a faithfully prolonged ideal?

REFERENCES

[1] É. CARTAN, Les Systèmes Différentiel Extérieurs et Leurs Applications Géométriques, Hermann, Paris, 1946.

[2] J.A. SCHOUTEN AND W.v.d. KULK, Pfaff's Problem and Its Generalizations, Clarendon Press, Oxford, 1949.

[3] R. HERMANN, Differential Geometry and the Calculus of Variations, Academic Press, New York, New York, 1968.

__________, Advances in Math. 1 (1965), 265-317.

__________, Lectures in Mathematical Physics, Vol. II, W. A. Benjamin, Reading, Mass., 1972.

_________, Geometry, Physics and Systems, Marcel Dekker, New York, N. Y., 1973.

_________, Interdisciplinary Mathematics, Vol. I - IX, Math Sci Press, 18 Gibbs Street, Brookline, Mass. 02146, 1973.

[4] W. ŚLEBODZIŃSKI, Exterior Forms and Their Applications, Polish Scientific Publishers, Warsaw, 1970.

[5] F.B. ESTABROOK, Comments on generalized Hamiltonian dynamics, Phys. Rev. D 8 (1973), 2740-2743.

[6] F.B. ESTABROOK AND H.D. WAHLQUIST, The geometric approach to sets of ordinary differential equations and Hamiltonian dynamics, SIAM Rev. 17 (1975), 201-220.

[7] B.K. HARRISON AND F.B. ESTABROOK, Geometric approach to invariance groups and solution of partial differential systems, J. Mathematical Phys. 12 (1971), 653-666.

[8] H.D. WAHLQUIST AND F.B. ESTABROOK, Prolongation structures of nonlinear evolution equations, J. Mathematical Phys. 16 (1975), 1-7.

[9] _________________________, Bäcklund transformation for solutions of the Korteweg-deVries equation, Phys. Rev. Lett. 31 (1973), 1386-1390.

BÄCKLUND TRANSFORMATION OF POTENTIALS OF THE KORTEWEG-DEVRIES EQUATION AND THE INTERACTION OF SOLITONS WITH CNOIDAL WAVES*

Hugo D. Wahlquist

Jet Propulsion Laboratory
California Institute of Technology
Pasadena, California 91103

I. INTRODUCTION

In a previous paper [1] a new set of conservation laws and potential functions associated with the Korteweg-deVries equation was derived. The definition of these potentials and the equations they satisfy are repeated in Section II of the present paper. One of the potential functions--denoted $y(x,t)$--is called a "pseudopotential" to emphasize that it cannot be written as a quadrature over the KdV solution. It was shown that $y(x,t)$ is closely related to the inverse scattering method for the KdV equation and also that the Bäcklund transformation of a solution, $u(x,t)$, of the KdV equation [2] is simply expressed as an algebraic function of $y(x,t)$.

In Section III of the present paper we consider the effect of a Bäcklund transformation, not just on $u(x,t)$, but also on the potential functions and on the pseudopotential $y(x,t)$ itself. General recursion relations are derived for the result of successive Bäcklund transformations. Finally, a closed form expression is obtained for the effect of any number of Bäcklund transformations applied to an arbitrary starting solution. This expression, involving determinants of pseudopotentials, constitutes a formal generalization of Hirota's result [3] for the multisoliton solutions.

In Section IV the general steady progressing wave solution of the KdV equation is written in terms of the Weierstrass elliptic function. Then explicit expressions for the potential functions corresponding to these steady progressing

*This paper presents the results of one phase of research carried out at the Jet Propulsion Laboratory, California Institute of Technology, under Contract No. NAS7-100, sponsored by the National Aeronautics and Space Administration.

waves are obtained in Section V. In Section VI these results are applied to the Bäcklund transformation of cnoidal waves from which an infinite hierarchy of new analytical solutions to the KdV equation can be generated. The solution resulting from a single Bäcklund transformation is analyzed in some detail and can be described as a superposition of phase-shifted cnoidal waves and a "modulated soliton."

II. POTENTIALS AND PSEUDOPOTENTIALS FOR THE KORTEWEG-DeVRIES EQUATION

We write the KdV equation in the form

$$u_t + p_x + 12uz = 0 \tag{1}$$

where we use the derivative variables

$$z = u_x, \quad p = z_x = u_{xx} \ . \tag{2}$$

In [1] a set of eight potential functions y_k, $k = 1,\ldots,8$, associated with this equation was obtained from the prolongation structure of the KdV equation. Two of these, y_1 and y_4, are trivial. It is useful, however, to introduce another pair of variables, ψ and ϕ, algebraically related to y_3 and y_8, so we will still have eight functions to deal with. To reserve numerical subscripts for later use we shall here adopt the following notation for the eight variables:

$$\begin{aligned} y_2 &= s(x,t) \ , & y_3 &= v(x,t) = -\ln\psi(x,t) \ , \\ y_5 &= \ln r(x,t) \ , & y_6 &= q(x,t) \ , \\ y_7 &= w(x,t) \ , & y_8 &= y(x,t) = \frac{\phi(x,t)}{\psi(x,t)} \ . \end{aligned} \tag{3}$$

If u, satisfying (1) and (2), is given, then each of these potential functions is defined by a pair of first-order equations in the following sequence:

$$w_x = -u , \quad w_t = p + 6u^2$$

(4a) $$q_x = -u^2 , \quad q_t = 8u^3 + 2up - z^2 ,$$

$$r_x = -wr , \quad r_t = (6q - z)r ,$$

$$y_x = -2u - y^2 + \lambda , \quad y_t = 4[(u + \lambda)(2u + y^2 - \lambda) + \frac{1}{2} p - zy] ,$$

(4b) $$v_x = -y , \quad v_t = 4(u + \lambda)y - 2z ,$$

$$s_x = -e^{2v} , \quad s_t = 4(u + \lambda)e^{2v} ,$$

where λ is an arbitrary constant. The cross-derivative integrability condition for each pair of equations is satisfied by virtue of the entire set together with (1) and (2). The equations for ψ and ϕ can be obtained by straightforward substitution with the result

$$\psi_x = \phi , \quad \psi_t = 2z\psi - 4(u + \lambda)\phi ,$$

(5)

$$\phi_x = (\lambda - 2u)\psi , \quad \phi_t = [4(u + \lambda)(2u - \lambda) + 2p]\psi - 2z\phi .$$

These are essentially the first-order linear equations which have been used to implement the inverse-scattering method for solving the KdV equation.

The simplest integral of the entire set of equations can be given by adopting the null KdV solution

(6) $$u(x,t) = z(x,t) = p(x,t) = 0 .$$

The potential functions w and q are then clearly constants, $w = w_0$, $q = q_0$, while the remaining variables become

$$r = r_0 e^{-w_0 x + 6q_0 t} ,$$

$$y = \lambda^{\frac{1}{2}} \tanh \theta ,$$

$$\psi = \psi_0 \cosh\theta\ ,$$

(7)
$$\phi = \psi_0\lambda^{\frac{1}{2}} \sinh\theta,$$

$$v = -\ln(\psi_0 \cosh\theta)\ ,$$

$$s = -\lambda^{-\frac{1}{2}}\psi_0^{-2} \tanh\theta + s_0\ ,$$

where $\theta \equiv \lambda^{\frac{1}{2}}(x - 4\lambda t) + \theta_0$.

III. BÄCKLUND TRANSFORMATIONS OF THE POTENTIALS

The Backlund transformation (B.T.) for solutions of the KdV equation [2] is rather clumsy if written directly for derivatives of the dependent variable $u(x,t)$. Using the potential function w, it takes the neater form

(8a)
$$\tilde{w}_x + w_x = -\lambda + (\tilde{w} - w)^2\ ,$$

(8b)
$$\tilde{w}_t + w_t = 4(\tilde{u}^2 + \tilde{u}u + u^2) + 2(\tilde{z} - z)(\tilde{w} - w)\ ,$$

where $\tilde{z} \equiv \tilde{u}_x$, $\tilde{p} \equiv \tilde{z}_x$, and λ again is an arbitrary constant. If we are given $w(x,t)$ such that $w_x = -u$, $w_t = p + 6u^2$ (which ensures that u is a solution of KdV), then the solution $\tilde{w}(x,t)$ of (8) satisfies

(9)
$$\tilde{w}_x = -\tilde{u}\ ,\quad \tilde{w}_t = \tilde{p} + 6\tilde{u}^2\ ,$$

guaranteeing that $\tilde{u}$ is another KdV solution.

Define the function W by

(10)
$$W \equiv w - \tilde{w}$$

Then (8a) can be rewritten in two other forms

(11)
$$\tilde{u} = -u - W^2 + \lambda$$

and

(12)
$$W_x = -2u - W^2 + \lambda\ .$$

It follows from differentiating the first of these and using the second, that

(13) $$\tilde{z} = -z + 2W(2u + W^2 - \lambda) \ .$$

When these results are inserted in (8b) we find

(14) $$W_t = 4[(u + \lambda)(2u + W^2 - \lambda) + \frac{1}{2} p - zW] \ .$$

A comparison of (12), (14), and (4) shows that W can be identified with the pseudopotential y, since they satisfy precisely the same pair of differential equations. Thus, given any $u(x,t)$, if we solve for the pseudopotential y, then we can accomplish a Bäcklund transformation simply by writing (cf. (10) and (11))

(15) $$\tilde{w} = w - y \ ,$$

and by differentiation

(16) $$\tilde{u} = - u - y^2 + \lambda \ ,$$

(17) $$\tilde{z} = - z + 2y(2u + y^2 - \lambda) \ ,$$

(18) $$\tilde{p} = - p + 4yz - 2(2u + y^2 - \lambda)(2u + 3y^2 - \lambda) \ .$$

In the following we shall continually be using notation equations (3) and the potential equations (4) and (5), together with (15)-(18). From (3) and (4), we have

(19) $$w = -(\ell n r)_x \ , \quad y = -v_x = (\ell n \psi)_x \ .$$

Thus, we can integrate (15) to find the Bäcklund transformation of the potential r

(20) $$\tilde{r} = r\psi \quad .$$

(A multiplicative integration function, say $f(t)$, could be included here, but can be taken equal to unity without loss of generality.) The time derivative of (20) gives

(21) $$\tilde{q} = q + \frac{1}{3} y^3 - \lambda y \quad .$$

With this result, we have obtained expressions for the transformation of all variables up to the pseudopotential y itself.

The transformation for y can be "derived" by assuming the property of permutation symmetry for consecutive Bäcklund transformations [4], [5]. Multiple B.T.'s require considering several values of the λ parameter, for which we shall use numerical subscripts, $\lambda_1, \lambda_2, \ldots, \lambda_m$, etc. We also need notation for the various functions obtained at each transformation step. The functions u, z, p, w, q, r, when subjected to N successive B.T.'s, will depend on N transformation parameters λ_m, $m = 1,\ldots,N$, and we shall indicate this dependence of the transformed functions by a sequence of subscripts. For instance, after $N = 3$ transformations,

$$\tilde{u}(x,t,\lambda_1,\lambda_2,\lambda_3) = u_{123} \,. \tag{22}$$

The pseudopotential y corresponding to the parameter λ_m will be denoted $y_{;m}$ and its x-equation will read

$$y_{;m,x} = -2u - (y_{;m})^2 + \lambda_m \,. \tag{23}$$

Through the B.T. this $y_{;m}$ leads to the transformed function u_m by (16)

$$u_m = -u - (y_{;m})^2 + \lambda_m \,. \tag{24}$$

To illustrate the subscript notation (22) for a sequence of B.T.'s, the fourth step of (23) and (24) would take the form

$$y_{123;m,x} = -2u_{123} - (y_{123;m})^2 + \lambda_m \tag{25}$$

and

$$u_{123m} = -u_{123} - (y_{123;m})^2 + \lambda_m \,. \tag{26}$$

The same notation will be used for the potentials ψ, ϕ, v, and s which depend on y, i.e.

$$y_{12;m} = \frac{\phi_{12;m}}{\psi_{12;m}} \,. \tag{27}$$

Consider a sequence of two B.T.'s, the first depending on λ_1 followed by a second depending on λ_2. The transformations of w and u will be

(28)
$$w_1 = w - y_{;1}\,, \quad u_1 = -u - y_{;1}^2 + \lambda_1\,,$$
$$w_{12} = w_1 - y_{1;2} = w - y_{;1} - y_{1;2}\,,$$
$$u_{12} = -u_1 - y_{1;2}^2 + \lambda_2 = u + y_{;1}^2 - \lambda_1 - y_{1;2}^2 + \lambda_2\,.$$

If the transformations are applied in reverse order, we again obtain (28) with the interchange $(1\leftrightarrows 2)$. Permutation symmetry, however, would require $w_{12} = w_{21}$ and $u_{12} = u_{21}$, and when these two equations are solved for $y_{1;2}$ $(y_{2;1})$, we find

(29)
$$y_{1;2} = \frac{\lambda_1 - \lambda_2}{y_{;1} - y_{;2}} - y_{;1}\,,$$

(30)
$$y_{2;1} = \frac{\lambda_2 - \lambda_1}{y_{;2} - y_{;1}} - y_{;2}\,.$$

Using (29) and inserting all previous results, one verifies that $y_{1;2}$ indeed satisfies the appropriate pseudopotential equations,

(31)
$$y_{1;2,x} = -2u_1 - y_{1;2}^2 + \lambda_2\,,$$
$$y_{1;2,t} = 4[(u_1 + \lambda_2)(2u_1 + y_{1;2}^2 - \lambda_2) + \tfrac{1}{2}p_1 - z_1 y_{1;2}]\,.$$

Since the starting solutions were unspecified, we can write these equations as a set of recursion relations for the pseudupotentials at each step of a sequence of transformations. Let μ_m, $m = 1,\dots,n$, denote the members of a set of n integers. Then we have

(32)
$$y_{\mu_1\dots\mu_{n-2}\mu_{n-1};\mu_n} = \frac{\lambda_{\mu_{n-1}} - \lambda_{\mu_n}}{y_{\mu_1\dots\mu_{n-2};\mu_{n-1}} - y_{\mu_1\dots\mu_{n-2};\mu_n}} - y_{\mu_1\dots\mu_{n-2};\mu_{n-1}}$$

together with

(33)
$$u_{\mu_1\dots\mu_n} = -u_{\mu_1\dots\mu_{n-1}} - (y_{\mu_1\dots\mu_{n-1};\mu_n})^2 + \lambda_{\mu_n}\,.$$

Thus, if the original pseudopotential $y(x,t,\lambda)$ can be found for arbitrary λ, we generate algebraically an infinite hierarchy of solutions. For example, with $y_{;1}, y_{;2}, y_{;3}$ we can construct $y_{1;2}$ and $y_{1;3}$ from (32), and then $y_{12;3}$ from the same equation. At each step, (33) gives the new KdV solution; viz.

$$
\begin{aligned}
u_1 &= - u - y_{;1}^2 + \lambda_1 , \\
u_{12} &= - u_1 - y_{1;2}^2 + \lambda_2 , \\
u_{123} &= - u_{12} - y_{12;3}^2 + \lambda_3 .
\end{aligned}
\tag{34}
$$

In Section II, (7), the general $y(x,t,\lambda)$ is given for the null KdV solution $u(x,t) = 0$. In this case the above process generates the infinite hierarchy of pure multisoliton solutions of the KdV equation.

To find the transformation of ψ, we define the temporary variable $f = \psi_{;1}\,\psi_{1;2}$ in order to rewrite (29) as

$$
\frac{f_x}{f} = \frac{\lambda_1 - \lambda_2}{y_{;1} - y_{;2}} . \tag{35}
$$

Differentiating this, using (23) for the derivatives of $y_{;1}$ and $y_{;2}$, and eliminating f with (35) gives

$$
\frac{f_{xx}}{f_x} = (\ell n \psi_{;1}\psi_{;2})_x , \tag{36}
$$

where (19) has been used to express the right side in terms of ψ's. Integrating yields

$$
f_x = \psi_{;1}\psi_{;2} , \tag{37}
$$

and putting this and f itself back in (35), we obtain*

* In Section V it will be shown that these ψ functions satisfy the well-known Schrödinger equation which is associated with the KdV equation [8]. The transformation formulas for ψ, (38)-(40), together with those for u in (34), which here have been derived from the Bäcklund transformation approach, can also be obtained from a study of the transformation properties of that Schrödinger equation [6].

$$(38)\qquad \psi_{1;2} = \psi_{;2}\left(\frac{y_{;1} - y_{;2}}{\lambda_1 - \lambda_2}\right) = \frac{1}{\psi_{;1}}\left(\frac{\phi_{;2}\psi_{;1} - \phi_{;1}\psi_{;2}}{\lambda_2 - \lambda_1}\right) .$$

This also provides $v_{1;2} = -\ln\psi_{1;2}$ and $\phi_{1;2} = \psi_{1;2}y_{1;2} = -\frac{\phi_{;1}}{\psi_{;1}}\psi_{1;2} + \psi_{;2}$.

Applying (38) for the next transformation step gives

$$(39)\qquad \psi_{12;3} = \psi_{1;3}\,\frac{y_{1;2} - y_{1;3}}{\lambda_2 - \lambda_3}$$

which can be expanded to

$$(40)\qquad \psi_{12;3} = \frac{1}{\psi_{;1}\psi_{1;2}}\;\frac{\phi_{;1}\psi_{;2}\psi_{;3}(\lambda_2-\lambda_3)+\phi_{;2}\psi_{;3}\psi_{;1}(\lambda_3-\lambda_1)+\phi_{;3}\psi_{;1}\psi_{;2}(\lambda_1-\lambda_2)}{(\lambda_1 - \lambda_2)(\lambda_2 - \lambda_3)(\lambda_3 - \lambda_1)} .$$

Comparing (38) and (40), we can begin to recognize the emerging pattern of determinants. Thus, let $\psi_{(N)}$ represent the potential with N subscripts, $(N) = 12\ldots N-1;N$; let $\phi_{;m}$, $\psi_{;m}$, $m = 1,2,\ldots,N$ represent the original untransformed potentials each depending on the single parameter λ_m. Define the square matrices, P_N and Λ_N, with components

$$(41)\qquad \begin{aligned} &(P_N)_{m,2p-1} = \psi_{;m}(\lambda_m)^{p-1} , \quad (P_N)_{m,2p} = \phi_{;m}(\lambda_m)^{p-1} \\ &(\Lambda_N)_{m,n} = (\lambda_m)^{n-1} , \quad m,n = 1,2,\ldots,N,\ p = 1,2,\ldots,[\tfrac{N+1}{2}] , \end{aligned}$$

where $[(N+1)/2]$ represents the greatest integer less than or equal to $(N+1)/2$. Then

$$(42)\qquad \psi_{(N)} = \left(\prod_{M=0}^{N-1}\psi_{(M)}\right)^{-1}\frac{|P_N|}{|\Lambda_N|} ,$$

where $\psi_{(0)} \equiv 1$.

For $N = 4$, for example, we have

$$P_4 = \begin{pmatrix} \psi_{;1} & \phi_{;1} & \lambda_1\psi_{;1} & \lambda_1\phi_{;1} \\ \psi_{;2} & \phi_{;2} & \lambda_2\psi_{;2} & \lambda_2\phi_{;3} \\ \psi_{;3} & \phi_{;3} & \lambda_3\psi_{;3} & \lambda_3\phi_{;3} \\ \psi_{;4} & \phi_{;4} & \lambda_4\psi_{;4} & \lambda_4\phi_{;4} \end{pmatrix} ,$$

$$\Lambda_4 = \begin{pmatrix} 1 & \lambda_1 & \lambda_1^2 & \lambda_1^3 \\ 1 & \lambda_2 & \lambda_2^2 & \lambda_2^3 \\ 1 & \lambda_3 & \lambda_3^2 & \lambda_3^3 \\ 1 & \lambda_4 & \lambda_4^2 & \lambda_4^3 \end{pmatrix} .$$

The transformation equation (20) for the potential r reveals a similar pattern, i.e.

(43)
$$r_1 = r\psi_{;1} ,$$
$$r_{12} = r_1\psi_{1;2} = r\psi_{;1}\psi_{1;2} ,$$

or for multiple transformations

(44)
$$r_{(N)} = r \prod_{M=0}^{N} \psi_{(M)} .$$

From (42) this can be rewritten as

(45)
$$r_{(N)} = r \frac{|P_N|}{|\Lambda_N|} .$$

Since $u_{(N)} = [\ell n\, r_{(N)}]_{xx}$, this gives a generalization of Hirota's form [3] for the multisoliton KdV solutions. Equation (45) applies for the infinite hierarchy of solutions generated by B.T.'s from <u>any</u> beginning solution of the KdV equation.

IV. STEADY PROGRESSING WAVE SOLUTIONS

The general steady progressing wave solutions of the KdV equation (1) can be written

$$(46) \qquad u(x,t) = k^2[\tfrac{1}{3} - P(\tau)] \ ,$$

where

$$(47) \qquad \tau = k(x - 4k^2t) + \tau_0 \ ,$$

k and τ_0 are arbitrary <u>complex</u> constants and $P(\tau)$ is the basic Weierstrass elliptic function. In general, this $u(x,t)$ is complex valued, but the interesting real solutions can be obtained for particular values of k and τ_0. We verify that (46) is a solution of (1); that (46) is the general steady progressing wave solutions requires the converse. Using primes to denote derivatives with respect to τ,

$$(48) \qquad \begin{aligned} u_t &= 4k^5P' \\ z \equiv u_x &= -k^3P' \ , \\ p \equiv z_x &= -k^4P'' \ , \end{aligned}$$

and inserting these into (1) gives

$$(49) \qquad P''' = 12PP'$$

which is the fundamental differential equation satisfied by all $P(\tau)$.

Detailed descriptions of the Weierstrass functions can be found in Abramowitz and Stegun [7,pp. 629-670] and we repeat here only those definitions and properties which are essential to the following analysis. The Weierstrass P-function, $P(\tau/\omega,\omega')$, is an even, doubly-periodic function (complex periods 2ω and $2\omega'$), having a double pole at $\tau = 0$; $P(\tau) - \tau^{-2}$ is analytic around the origin and vanishes at $\tau = 0$. The integrals of (49) are commonly written

$$(50) \qquad \begin{aligned} P'' &= 6P^2(\tau) - \tfrac{1}{2}g_2 \ , \\ (P')^2 &= 4P^3(\tau) - g_2P(\tau) - g_3 \\ &= 4(P - e_1)(P - e_2)(P - e_3) \ , \end{aligned}$$

where the parameters are restricted by

$$g_2 = 2(e_1^2 + e_2^2 + e_3^2) \; ,$$

(51) $$g_3 = 4e_1e_2e_3 \; ,$$

$$e_1 + e_2 + e_3 = 0 \; .$$

Using the notation

(52) $$\omega_1 = \omega \; , \; \omega_2 = \omega + \omega' \; , \; \omega_3 = \omega' \; ,$$

one has the periodicity relations and special values ($j = 1,2,3$)

(53)
$$\begin{aligned} P(\tau + 2\omega_j) &= P(\tau) \; , \\ P'(\tau + 2\omega_j) &= P'(\tau) \; , \\ P(\omega_j) &= e_j \; , \\ P'(\omega_j) &= 0 \; . \end{aligned}$$

The relation of $P(\tau)$ to the Jacobi elliptic functions, sn and cn, can be expressed by [7,p. 659]

(54)
$$\begin{aligned} P(\tau) &= e_3 + \frac{\gamma^2}{\mathrm{sn}^2(\gamma\tau|m)} = e_3 + m\gamma^2\mathrm{sn}^2(\gamma\tau - iK'|m) \\ &= e_2 - (e_2 - e_3)\mathrm{cn}^2(\gamma\tau - iK'|m) \end{aligned}$$

where

(55)
$$\begin{aligned} &e_1 = \tfrac{1}{3}\gamma^2(2-m), \; e_2 = \tfrac{1}{3}\gamma^2(2m-1), \; e_3 = -\tfrac{1}{3}\gamma^2(m+1) \; , \\ &\gamma^2 = e_1-e_3 \; , \; m = \frac{e_2-e_3}{e_1-e_3} \; , \; K' = K(1-m) \; , \end{aligned}$$

and K is the complete elliptic integral of the first kind. Thus, for various choices of the parameters k and τ_0, one obtains the real, regular, cnoidal-wave solutions, as well as a multiple of complex and real singular waves.

For example, if $e_2 = e_3$, we have

(56) $$e_1 = \tfrac{2}{3}\gamma^2, \; e_2 = e_3 = -\tfrac{1}{3}\gamma^2 \; , \; m = 0 \; ,$$

and

$$(57)\qquad P(\tau) = \gamma^2[-\frac{1}{3} + \csc^2(\gamma\tau)] \ .$$

giving the singular trigonometric waves. The limit $\gamma \to 0$ here gives the singular KdV solution

$$(58)\qquad u = k^2(\frac{1}{3} - \frac{1}{\tau^2}) \ .$$

If $e_1 = e_2$, we have

$$(59)\qquad e_1 = e_2 = \frac{1}{3}\gamma^2, \quad e_3 = -\frac{2}{3}\gamma^2 \ , \quad m = 1 \ ,$$

and

$$(60)\qquad P(\tau) = \gamma^2\left[\frac{1}{3} + \frac{1}{\sinh^2(\gamma\tau)}\right] = \gamma^2 \left[\frac{1}{3} - \mathrm{sech}^2(\gamma\tau - i\,\frac{\pi}{2})\right] \ ,$$

so that for appropriate choices of the parameters, we obtain the real 1-soliton solutions, regular and singular. The soliton solutions which vanish asymptotically $(u \to 0;\ |x| \to \infty)$ are obtained for $\gamma = 1$.

V. POTENTIALS OF THE STEADY PROGRESSING WAVE SOLUTIONS

To obtain explicit expressions for the potentials and pseudopotentials of the steady progressing wave solutions, we need the related Weierstrass ζ- and σ-functions defined by

$$(61)\qquad \zeta'(\tau) = -P(\tau) \ , \quad \frac{\sigma'(\tau)}{\sigma(\tau)} = \zeta(\tau) \ .$$

These are not "elliptic" functions, since they are not strictly periodic. The ζ-function is odd, $\zeta(-\tau) = -\zeta(\tau)$; $\zeta(\tau) - \tau^{-1}$ vanishes at $\tau = 0$, and is analytic in the neighborhood of the origin; the σ-function is an entire function which vanishes at $\tau = 0$. To simplify the writing of some expressions, it is convenient to introduce the function $\eta(\tau)$

$$(63)\qquad \sigma(\tau) = e^{\eta(\tau)} \ ,$$

so

$$(64)\qquad \eta'(\tau) = \zeta(\tau) \ .$$

The basic addition and multiplication formulas needed for these functions are

$$P(\tau_1 + \tau_2) + P(\tau_1) + P(\tau_2) = Q^2(\tau_1,\tau_2) ,$$

(65)

$$\zeta(\tau_1 + \tau_2) - \zeta(\tau_1) - \zeta(\tau_2) = Q(\tau_1,\tau_2) ,$$

where Q is defined by

(66) $$Q(\tau_1,\tau_2) \equiv \frac{1}{2}\frac{P'(\tau_1) - P'(\tau_2)}{P(\tau_1) - P(\tau_2)} .$$

The expressions for w, q, and r can be obtained directly from (46) by quadrature as

(67) $$w = - k[\frac{1}{3}\tau + \zeta(\tau) - \frac{1}{2}k^3(g_2 - \frac{4}{3})t] ,$$

(68) $$q = -k^3[\frac{1}{6}P'(\tau) + \frac{2}{3}\zeta(\tau) + \frac{1}{12}(g_2 + \frac{4}{3})\tau - k^3(g_3 + \frac{1}{3}g_2 - \frac{4}{27})t] ,$$

(69) $$r = \exp\{\frac{1}{6}\tau^2 + \eta(\tau) - \frac{1}{2}k^3(g_2 - \frac{4}{3})t\tau + 3k^6(g_3 + \frac{8}{27})t^2\}.$$

Using (46), (48), (50), (61), and (64) one verifies that the above expressions satisfy the potential equations, (4), for w, q, and r.

To obtain the pseudopotential y, we use (19)

(70) $$y = \frac{\psi_x}{\psi}$$

to convert the Riccati equation

(71) $$y_x = - 2u - y^2 + \lambda$$

to the well-known linear second-order equation associated with the KdV solutions u [8]

(72) $$\psi_{xx} + (2u - \lambda)\psi = 0 .$$

For the steady progressing wave solution (46), this becomes

(73) $$\psi'' - [(\frac{\lambda}{k^2} - \frac{2}{3}) + 2P(\tau)]\psi = 0 ,$$

where we are again using τ derivatives. Equation (73) is a standard form of the Lamé equation which is treated in Ince [9,p. 378]. Two generally distinct solutions, as given by Ince, are

$$\psi_1 = \frac{\sigma(\tau+a)}{\sigma(\tau)} e^{-\tau\zeta(a)} \quad , \quad \psi_2 = \frac{\sigma(\tau-a)}{\sigma(\tau)} e^{\tau\zeta(a)} \quad , \tag{74}$$

where the constant a is determined by

$$P(a) = \frac{\lambda}{k^2} - \frac{2}{3} \quad . \tag{75}$$

If λ is such that

$$P(a) = e_j \quad , \quad j = 1,2,3, \tag{76}$$

then $a = \omega_j$, and ψ_1 and ψ_2 are no longer independent. The independent solutions for such special cases can be written [9,p. 380]

$$\begin{aligned} \psi_1^j &= [P(\tau) - e_j]^{\frac{1}{2}} \quad , \\ \psi_2^j &= [P(\tau) - e_j]^{\frac{1}{2}}[\zeta(\tau + \omega_j) + e_j\tau] \quad . \end{aligned} \tag{77}$$

For the present, we assume (76) is not satisfied, so that the general solution of (73) for the pseudopotential ψ (with arbitrary λ) can be taken from (74) as

$$\psi = e^{\theta_1} + e^{\theta_2} \tag{78}$$

where

$$\begin{aligned} \theta_1 &= \eta(\tau + a) - \eta(\tau) - \zeta(a)\tau + k^3 f_1(t) \quad , \\ \theta_2 &= \eta(\tau - a) - \eta(\tau) + \zeta(a)\tau + k^3 f_2(t) \quad , \end{aligned} \tag{79}$$

and $f_1(t)$, $f_2(t)$ are integration functions which remain to be determined.

From (5) we have

$$\phi = \psi_x = k\psi' = k[e^{\theta_1}Q(\tau,a) + e^{\theta_2}Q(\tau,-a)] \tag{80}$$

where (65) has been used to write

$$\theta_1' = \zeta(\tau + a) - \zeta(\tau) - \zeta(a) = Q(\tau,a) \ ,$$

(81)

$$\theta_2' = \zeta(\tau - a) - \zeta(\tau) - \zeta(-a) = Q(\tau,-a) \ .$$

Now substituting ψ and ϕ into the remaining relations of (5), one finds

$$\dot{f}_1 = -2\, P'(a)$$

(82)

$$\dot{f}_2 = 2\, P'(a) \ .$$

Thus, ψ and ϕ are given by (78) and (80) with

$$\theta_1 = \eta(\tau + a) - \eta(\tau) - \zeta(a)\tau - 2k^3 P'(a)t + \beta_1 \ ,$$

(83)

$$\theta_2 = \eta(\tau - a) - \eta(\tau) + \zeta(a)\tau + 2k^3 P'(a)t + \beta_2 \ ,$$

where β_1 and β_2 are arbitrary constants. Using these results in (45) at the end of Section III will give the infinite hierarchy of solutions generated by recursive Bäcklund transformations of any steady progressing wave solution.

The form of the pseudopotential y given by

(84) $$y = \frac{\phi}{\psi} = k\, \frac{e^{\theta_1} Q(\tau,a) + e^{\theta_2} Q(\tau,-a)}{e^{\theta_1} + e^{\theta_2}}$$

is more convenient for analytically expressing the result of a transformation of the KdV solution directly. Before continuing, we note the effect of using just <u>one</u> of the two independent solutions to (73). For example, by letting $\beta_2 \to -\infty$ in (83), we have $\theta_2 \to -\infty$, and

(85) $$y = kQ(\tau,a) \ .$$

Thus, the transformed solution

(86) $$\tilde{u} = -u - y^2 + \lambda$$

becomes

$$\tilde{u} = -k^2[\tfrac{1}{3} - P(\tau)] - k^2[P(\tau + a) + P(\tau) + P(a)] + k^2[\tfrac{2}{3} + P(a)] \tag{87}$$

where the addition theorem of (65) has been used in the second term. Collecting terms in (87) we have

$$\tilde{u} = k^2[\tfrac{1}{3} - P(\tau + a)] , \tag{88}$$

so the only result is a translation of the original solution. The second solution alone, of course, gives

$$\tilde{u} = k^2[\tfrac{1}{3} - P(\tau - a)] . \tag{89}$$

In general, these are translations in the complex plane so they are not necessarily equivalent merely to phase shifts. They may, for instance, also convert real regular solutions into real singular solutions or vice versa.

The general expression for y can be simplified by defining a variable θ as

$$2\theta \equiv \theta_1 - \theta_2 = \eta(\tau + a) - \eta(\tau - a) - 2\zeta(a)\tau - 4k^3P'(a)t + \beta \tag{90}$$

where $\beta \equiv \beta_1 - \beta_2$. Then (84) becomes

$$y = \frac{k}{2} \operatorname{sech} \theta[e^{\theta}Q(\tau,a) + e^{-\theta}Q(\tau,-a)] , \tag{91}$$

and using the definition of $Q(\tau,a)$, (66), we get

$$y = \frac{k}{2} \frac{[P'(\tau) - \tanh(\theta)P'(a)]}{[P(\tau) - P(a)]} , \tag{92}$$

which can also be written in terms of the KdV variables directly

$$y = \frac{1}{2} \frac{[z(\tau) - \tanh(\theta)z(a)]}{[u(\tau) - u(a)]} \tag{93}$$

The variable θ, defined in (90), can be expressed in terms of more familiar elliptic functions by first differentiating it with respect to τ and using (64), (65), and (66). The result is

(94) $$\theta' = -\frac{1}{2}\ \frac{P'(a)}{P(\tau)-P(a)}$$

or from (54)

(95) $$\theta' = \frac{P'(a)}{2\gamma^2}\ \mathrm{sn}^2(\gamma a)[1 - m\ \mathrm{sn}^2(\gamma a)\mathrm{sn}^2(\gamma\tau - iK')]^{-1} .$$

The modulus parameter of all these elliptic functions is understood to be m, and the constant $P'(a)$ is given by

(96) $$P'(a) = -2\gamma^3\ \frac{\mathrm{cn}(\gamma a)\mathrm{dn}(\gamma a)}{\mathrm{sn}^2(\gamma a)} ,$$

where dn is another Jacobi elliptic function. Let $\gamma\tau_0 = \xi_0 + iK'$ where we assume ξ_0 real and introduce the real variable

(97) $$\xi = \gamma\tau - iK' = \gamma k(x - 4k^2t) + \xi_0 ,$$

and the parameter

(98) $$n = m\ \mathrm{sn}^2(\gamma a) .$$

Reintegrating (95), we have

(99) $$\theta = P'(a)\left(\frac{\mathrm{sn}^2(\gamma a)}{2\gamma^3}\int_0^{\xi}[1 - n\ \mathrm{sn}^2(\nu)]^{-1}d\nu - 2k^3t + \beta_0\right)$$

where β_0 is a new constant. The integral in (99) is a standard form for Π, the elliptic integral of the third kind [7,pp. 590], so that finally

(100) $$\theta = -N[n\Pi(n;\xi|m) - 4mk^3\gamma^3t + b]$$

where

(101)
$$N = [n^{-3}(1 - n)(m - n)]^{1/2} ,$$
$$b = 2m\gamma^3\beta_0 .$$

VI. BÄCKLUND TRANSFORMATION OF A CNOIDAL WAVE

We use the parameters $\{m, \gamma, n\}$ defined in (55) and (98), and henceforth assume $0 \le n \le m \le 1$ to guarantee real, non-singular solutions. Equation (54) then gives

$$P(\xi) = -\frac{1}{3}\gamma^2(m+1) + m\gamma^2 \mathrm{sn}^2(\xi) ,$$

(102)

$$P(a) = -\frac{1}{3}\gamma^2(m+1) + \frac{\gamma^2}{\mathrm{sn}^2(\gamma a)} = -\frac{1}{3}\gamma^2(m+1) + \frac{m\gamma^2}{n} ,$$

where e_3 has been replaced using (55). The expressions for u and λ, (46) and (75), become

$$u = \frac{1}{3}k^2[1 + \gamma^2(1+m) - 3m\gamma^2 \mathrm{sn}^2(\xi)] , \tag{103}$$

$$\lambda = \frac{1}{3}k^2[2 - \gamma^2(1+m-3\frac{m}{n})] , \tag{104}$$

and the expression for y (93) becomes

$$y = -k\gamma n[\mathrm{sn}(\xi)\mathrm{cn}(\xi)\mathrm{dn}(\xi) + N \tanh\theta][1 - n\mathrm{sn}^2(\xi)]^{-1} . \tag{105}$$

Then the transformed KdV solution $\tilde{u} = -u - y^2 + \lambda$ can be written

$$\tilde{u} = \tilde{u}_p + \frac{k^2\gamma^2n^2N^2}{[1-n\ \mathrm{sn}^2(\xi)]^2}\mathrm{sech}^2\theta - \frac{2k^2\gamma^2n^2N\ \mathrm{sn}(\xi)\mathrm{cn}(\xi)\mathrm{dn}(\xi)}{[1-n\ \mathrm{sn}^2(\xi)]^2}\tanh\theta, \tag{106}$$

where $\tilde{u}_p$ is the periodic function

$$\tilde{u}_p = \frac{1}{3}k^2\{1 + \gamma^2(1+m) - \frac{3}{2}m\gamma^2[\mathrm{sn}^2(\xi+\gamma a)+\mathrm{sn}^2(\xi-\gamma a)]\} \tag{107}$$

$$= \frac{1}{2}\{u(\xi + \gamma a) + u(\xi - \gamma a)\} .$$

To obtain this we have applied the addition theorem for sn functions which for our purpose can be written

$$\mathrm{sn}(\xi \pm \gamma a) = \left(\frac{n}{m}\right)^{\frac{1}{2}} \left[\frac{\mathrm{cn}(\xi)\mathrm{dn}(\xi) \pm nN\ \mathrm{sn}(\xi)}{1-n\ \mathrm{sn}^2(\xi)}\right] . \tag{108}$$

Thus $\tilde{u}$ given by (106) appears as a superposition of three types of waves: (1) A purely periodic elliptic wave, $\tilde{u}_p$, consisting of the sum of two components, each of which is one-half the original solution phase-shifted by $\pm\gamma a$. The phase velocity, v_p, of these waves is given by

$$v_p = -\frac{\xi_t}{\xi_x} = 4k^2 . \tag{109}$$

(2) The intrinsically positive term whose form suggests calling it a "modulated soliton." The phase velocity, v_s, of the soliton is given by

$$v_s = -\frac{\theta_t}{\theta_x} = v_p\{1 + \frac{m\gamma^2}{n}[1 - n\,\mathrm{sn}^2(\xi)]\} \tag{110}$$

$$= 4(\lambda + u) .$$

This velocity varies periodically but is always larger than v_p, so the soliton propagates across the associated elliptic wave. Note that at the zeroes of the original solution (i.e. where $u = 0$), this soliton velocity agrees with the expected velocity for an isolated soliton in a pure multisoliton solution.

(3) An oscillatory wave which vanishes at the soliton peak and grows on both sides to finite asymptotic amplitude. Far from the soliton peak ($|\theta| \to \infty$) this wave combines with $\tilde{u}_p$ to produce the asymptotic solutions

$$\tilde{u}_{as} = u(\xi \pm \gamma a) , \tag{111}$$

corresponding to simple phase shifts of the original solution.

The solutions corresponding to limiting values of the parameter n ($n = a = 0$ and $n = m$ or $\gamma a = K(m)$) are periodic, since in both cases $\tilde{u} = \tilde{u}_p$. This follows for $n \to 0$ from the fact that θ diverges and $n^2 N \to 0$, while for $n = m$ it results immediately from $N = 0$. Thus, in the first case

$$\tilde{u}(n = 0) = u(\xi) \tag{112}$$

and there is no transformation. In the second case

$$\tilde{u}(n = m) = u(\xi + K(m)) . \tag{113}$$

which is simply a phase shift of the original solution by one-half wavelength.

In the special case $m = 1$, all the Jacobi elliptic functions become the hyperbolic functions, e.g. $\mathrm{sn}(\xi|1) = \tanh \xi$. Then the original solution is the solution

(114) $$u = \frac{1}{3} k^2[1 - \gamma^2 + 3\gamma^2 \operatorname{sech}^2 \xi] ,$$

the transformation parameters being given by

(115) $$\begin{aligned} n &= \tanh^2(\gamma a) , \\ N &= n^{-3/2}(1 - n) , \\ \lambda &= k^2[\tfrac{2}{3}(1 - \gamma^2) + \frac{\gamma^2}{n}] , \end{aligned}$$

and the elliptic integral is

(116) $$\Pi(n;\xi|1) = \frac{1}{(n-1)} [\xi - n^{1/2}\tanh^{-1}(n^{1/2}\tanh \xi)] .$$

Defining parameters k' and γ' by

(117) $$k'\gamma' \equiv k\gamma n^{-\frac{1}{2}} , \quad k'^2(1 - \gamma'^2) \equiv k^2(1 - \gamma^2) ,$$

and the variable

(118) $$\xi' = k'\gamma'(x - 4k'^2 t) + n^{-3/2}(1 - n)b + n^{-\frac{1}{2}}\xi_0 ,$$

we can write

(119) $$\theta = \tanh^{-1}(n^{\frac{1}{2}}\tanh \xi) - \xi'$$

or

(120) $$\tanh \theta = \frac{n^{\frac{1}{2}}\tanh \xi - \tanh \xi'}{1 - n^{\frac{1}{2}}\tanh \xi \tanh \xi'} .$$

Thus, $\tilde{u}(\xi,\xi')$ is clearly just a 2-soliton solution which, however, does not vanish asymptotically. In the limits we have, for $|\xi'| \to \infty$,

(121) $$\tilde{u}(\xi) = \frac{1}{3} k^2[1 - \gamma^2 + 3\gamma^2\operatorname{sech}^2(\xi \pm \gamma a)]$$

and for $|\xi| \to \infty$,

(122) $$\tilde{u}(\xi') = \frac{1}{3} k'^2[1 - \gamma'^2 + 3\gamma'^2\operatorname{sech}^2(\xi' \pm \gamma a)] .$$

REFERENCES

[1] H.D. WAHLQUIST AND F.B. ESTABROOK, Prolongation structures of nonlinear evolution equations, J. Mathematical Phys. 16 (1975), 1-7.

[2] ________________, Bäcklund transformation for solutions of the Korteweg-deVries equation, Phys. Rev. Lett. 31 (1973), 1386-1390.

[3] R. HIROTA, Exact solution of the Korteweg-deVries equation for multiple collisions of solitons, Phys. Rev. Lett. 27 (1971), 1192-1194.

[4] L.P. EISENHART, A Treatise on the Differential Geometry of Curves and Surfaces, Dover Publications, New York, N. Y., 1960, 280-290.

[5] G.L. LAMB, JR., Analytical descriptions of ultrashort optical pulse propagation in a resonant medium, Rev. Modern Phys. 43 (1971), 99-124.

[6] L.D. FADDEYEV, The inverse problem in the quantum theory of scattering, J. Mathematical Phys. 4 (1963), 72-104. (Translated from the Russian by B. Seckler).

[7] M. ABRAMOWITZ AND I.A. STEGUN, Handbook of Mathematical Functions, National Bureau of Standards, U.S. Govt. Printing Office, Washington, D.C., 1964.

[8] C.S. GARDNER, J.M. GREENE, M.D. KRUSKAL, AND R.M. MIURA, Method for solving the Korteweg-deVries equation, Phys. Rev. Lett. 19 (1967), 1095-1097; Korteweg-deVries equation and generalizations. VI. Methods for exact solution, Comm. Pure Appl. Math. 27 (1974), 97-133.

[9] E.L. INCE, Ordinary Differential Equations, Dover Publications, New York, N.Y., 1944.

PSEUDOPOTENTIALS AND THEIR APPLICATIONS

James P. Corones and Frank J. Testa

Department of Mathematics
Iowa State University
Ames, Iowa 50010

I. INTRODUCTION

Several months ago Wahlquist and Estabrook [1] introduced the ideas of pseudopotentials and prolongation structures into the study of nonlinear partial differential equations. Their study of the Korteweg-deVries (KdV) equation suggests that the concept of a pseudopotential should be explored with the aim of deriving Bäcklund transformations, conservation laws, and associated linear problems from this single well-defined mathematical object. The remarks that follow are the first stage of such an exploration.

The point of view adopted is frankly computational, the aim being to produce examples which will act as guides for reasonable conjectures concerning a class or classes of equations. Since nearly any computation of the type presented below is rather lengthy, an attempt has been made to sharpen each question asked of the formalism in order to minimize the work necessary to produce an answer.

One of the difficulties with the original derivation of the prolongation structures is that it leads to a large number of quantities, many redundant, from which only a few selected objects are of "interest". If consideration is being given to an equation which has not previously been treated it is useful to have an *a priori* idea of which quantities are interesting. For example, if a search is being made for first-order linear eigenvalue and time evolution problems associated with a given equation, it is useful to realize that any eigenvalue problem of this type is defined via a linear pseudopotential (see below). Such objects can be directly and exhaustively computed without having to compute additional quantities.

Since the point of view adopted here is computational no general discussion of the *differential forms approach* used in [1] to produce pseudopotentials is needed. The differential forms approach is quickly sketched and it is pointed out how pseudopotentials can be computed by classical methods. Below this is called the direct approach. The relationship between the differential forms and classical approaches is not completely clear. This is due, at least, to the large number of (perhaps equivalent) ways an equation can be represented by a set of forms. The point being that what equivalence means with regard to computing pseudopotentials is obscure. However, what is empirically clear from [1], some yet unpublished work of Wahlquist and Estabrook, and the computations given below is that the concept of pseudopotentials can be taken as central in discussions of equations with soliton solutions. Why this should be true is an open question. One objective of this work is to motivate a search for its answer.

II. DIFFERENTIAL FORMS APPROACH AND DIRECT APPROACH

The result of the Wahlquist-Estabrook prolongation argument is the existence of certain exact 1-forms with a given partial differential equation as a constraint. More precisely, consider the function $u \equiv u(x,y)$ and let z^μ denote the set containing x, y, u, and all partial derivatives of u up through order μ,

$$z^\mu \equiv \{x,y,u,u_x,u_y,u_{xy},\dots\} . \tag{1}$$

Suppose $u(x,y)$ satisfies the nth-order nonlinear partial differential equation

$$N(z^n) = 0. \tag{2}$$

The Wahlquist-Estabrook approach provides a means of constructing exact 1-forms

$$dq^k = F^k(z^\mu,q)dx + G^k(z^\mu,q)dy, \quad k = 1,\dots,n, \tag{3}$$

where q denotes the set $\{q^1,q^2,\dots,q^n\}$, n arbitrary, clearly, since

$$q^k_x = F^k(z^\mu,q) \quad , \quad q^k_y = G^k(z^\mu,q) , \tag{4}$$

the necessary and sufficient conditions for exactness of (3) are given by

$$F^k_y(z^\mu,q) = G^k_x(z^\mu,q) . \tag{5}$$

The set of exact 1-forms given by (3) is the set of all pseudopotentials associated with (2), i.e. the set of all pseudopotentials associated with a given partial differential equation (2) is the set of all 1-forms (3) which are exact subject to the restriction that (2) is satisfied.

The Wahlquist and Estabrook method for determining functions F^k and G^k satisfying (5) and concomitant with (2), consists of first constructing a set of 2-forms $\{\alpha_\ell\}$ on the basis set $\{dz^\mu \wedge dx^\mu\}$ where the forms are null when z^μ is restricted to the solution manifold of (2),

(6) $$\tilde{\alpha}_\ell = 0,$$

and furthermore, they are closed under exterior differentiation,

(7) $$d\alpha_\ell = \sum \xi^\ell_j \wedge \alpha_j,$$

where ξ^ℓ_j is some set of 1-forms. Next, the primitive set containing z^μ is prolonged to include the variables q^k, $k = 1,\dots,n$, by introducing the 1-forms

(8) $$\omega_k = -dq^k + F^k(z^\mu,q)dx + G^k(z^\mu,q)dy, \quad k = 1,\dots,n.$$

These 1-forms are assumed to satisfy the closure relation

(9) $$d\omega_k = \sum_\ell f^k_\ell \alpha_\ell + \sum_i \eta^k_i \wedge \omega_i,$$

where η^k_i is some set of 1-forms, which lead to an overdetermined set of partial differential equations in F^k and G^k involving bilinear terms of the form

(10) $$\sum_\ell (F^k{}_{q^\ell} G^\ell - F^\ell G^k{}_{q^\ell})$$

The dependence of F^k and G^k on the primitive variables z^μ is then sought and, if obtained, yields a system of commutator type equations for the q^k dependence by virtue of the terms given in (10). The set of bilinear partial differential equations which the F^k and G^k satisfy will be called the prolongation structure associated with the original equation (2). When all terms of the form (10) are zero the prolongation structure is called abelian. Although the question of existence of nontrivial solutions of the prolongation structure may be studied by direct classical analysis, elegant attempts at producing particular solutions, based on Lie algebraic methods, have been proposed [1].

After determining functions F^k and G^k satisfying (8) and (9), it is required that the 1-forms ω_k be null when z^μ is restricted to the solution manifold of (2)

$$\tilde{\omega}_k = 0, \tag{11}$$

resulting in exact 1-forms (3), with $q(x,y)$ satisfying (4). That the functions F^k and G^k determined from (9) satisfy (5) follows from the consistency of (6) and (11) with (9). There is, of course, another approach to the problem of finding solutions of (3) subject to (2). The integrability conditions can be implemented directly, the z^μ still being treated as independent variables. This is the classical or what will be called the direct approach. It is preferable to the differential forms approach since it is usually computationally simpler and unambiguously exhausts all possible pseudopotentials.

III. SPECIALIZED FEATURES OF THE PROLONGATION STRUCTURE AND BÄCKLUND TRANSFORMATIONS

Consider now some specialized features of the full prolongation structure defined by (4) and (5). Observe first that the functions F^k and G^k are determined only up to an arbitrary transformation among the pseudopotentials q^k since the nonsingular transformation

$$\hat{q}^k \equiv \phi^k(q) \;,\; |\phi^k_\ell(q)| \neq 0,\; \phi^k_\ell(q) \equiv \frac{\partial \phi^k}{\partial q^\ell} \tag{12}$$

applied to (4) and (5) produces the equations

$$\hat{q}^k_x = \sum_\ell \phi^k_\ell F^\ell \equiv \hat{F}^k \;,\; \hat{q}^k_y = \sum_\ell \phi^k_\ell G^\ell \equiv \hat{G}^k, \tag{13}$$

with $\hat{F}^k$ and $\hat{G}^k$ satisfying (5). Because of the trivial nature of the $\hat{q}$ dependence of $\hat{F}^k$ and $\hat{G}^k$ through ϕ^k in (13), without loss of generality, it may be assumed that any such dependence has been removed by a transformation of the type

(12), thereby simplifying the calculation of F^k and G^k by classical methods.

In the special case where F^k and G^k are independent of q, the exact 1-forms in (3) reduce to a statement of a conservation law associated with (2), where $q(x,y)$ is the corresponding potential. This is clear from inspection of (3) when neither F^k nor G^k depend on any of the q's. This is a case of an _abelian_ prolongation structure. How higher conservation laws can be obtained in this way will be discussed below.

Another important specialization of the prolongation structure defined by (3), (4) and (5) is the case where $n = 1$, yielding the equations

(15) $$q_x = F(z^\mu,q) \ , \quad q_y = G(z^\mu,q) \ ,$$

subject to the integrability condition

(16) $$F_y(z^\mu,q) = G_x(z^\mu,q) \ .$$

These equations determine exact 1-forms of the type

(17) $$dq = F(z^\mu,q)dx + G(z^\mu,q)dy \ ,$$

for a pseudopotential $q(x,y)$, referred to as a pseudopotential _of the first kind_. Such pseudopotentials are of particular interest since the commutator type equations for the q-dependence of F and G obtained from (16) are simple ordinary differential equations, often admitting closed solutions. The general solution to (16) for F and G then determines via (15) the complete family of pseudopotentials of the first kind associated with a specific primitive set z^μ.

In addition it appears that pseudopotentials of the first kind can be used to obtain Bäcklund transformations, linear problems, and therefore higher conservation laws.

A lucid example of the calculation of pseudopotentials of the first kind is found by studying the equation

(18) $$u_{xy} = g(u),$$

with the associated primitive set z^μ taken to be of the form

(19) $$z^\mu \equiv \{u, z \equiv u_x, p \equiv u_y\} .$$

Straightforward analysis of (16) with (18) and (19) gives solutions for F and G given by

(20) $$F = z + I(q-\lambda u), \ G = \lambda p + J(q-u),$$

where the freedom of (12) has been used and I and J are disposable functions restricted by the condition

(21) $$J'(q-u)I(q-\lambda u) - I'(q-\lambda u)J(q-u) = (1-\lambda)g(u).$$

From (21) it follows that

(22) $$I'' = \alpha I, \ J'' = \alpha J, \ g'' = 2\alpha(1-\lambda)^2 g,$$

where α is a separation constant and $\lambda \neq 1$ is assumed thereby restricting the form of $g(u)$.* This fact demonstrates the sensitivity of the existence of pseudopotentials to the structure of (2).

Given the general pseudopotential of the first kind, defined by (17) and (20), one can, again following the lead of [1], look for a solution of (18) in the form

(23) $$u' = u'(z^\mu, q)$$

where $u \in z^\mu$ is a given solution to (18). If (23) is solved for q and the result is substituted into (20) for, say, $g(u) = \sin u$, the standard form of the Bäcklund transformation results. This appears to be a new approach to obtaining Bäcklund transformations. The introduction and subsequent removal of the function q has no role in the classical approach. It is not yet clear if this method is

* This condition has previously been derived by Kruskal in connection with the infinite number of conservation laws associated with (18), R.M. Miura, private communication. Also, see [2].

equivalent to the more usual (Clairin's) approach. However the conditions on $g(u)$ in (22) are the same as those in [2] for the existence of a Bäcklund transformation and are obtained here with considerably less labor.

Consider now the full prolongation structure of the KdV equation, restricted to the case where one pseudopotential is present, i.e. consider the pseudopotential of the first kind associated with the KdV equation. We learn from [1] that

$$F = 2x_1(q) + 2ux_2(q) + 3u^2x_3(q)$$

(24)

$$G = -2(p + 6u^2)x_2(q) + 3(z^2-8u^3-2up)x_3(q) + 8x_4(q) + 8ux_5(q) + 4u^2x_6(q) + 4zx_7(q),$$

$$[x_1,x_3] = [x_2,x_3] = x_1,x_4] = [x_2,x_6] = 0,$$

(25)

$$[x_1,x_2] = -x_7, \quad [x_1,x_7] = x_5, \quad [x_2x_7] = x_6$$

$$[x_1,x_5] + [x_2,x_4] = 0, \quad [x_3,x_4] + [x_1,x_6] + x_7 = 0,$$

where $z = u_x$, $p = u_{xx}$, and

$$[x_i,x_j] \equiv \frac{\partial x_i}{\partial q} x_j - x_i \frac{\partial x_j}{\partial q} .$$

Since only a single independent variable, q, is present if $[x_i,x_j] = 0$, integration yields $x_i = \alpha x_j$, α constant. Using this fact, if nonabelian prolongation structures are sought it is necessary to take x_1, x_2, and $[x_1,x_2]$ all not zero. From this it follows that $x_3 = 0$ and that the other x_j can be expressed in terms of, say, x_2, hence

$$x_1 = \frac{1}{2}\{-\alpha + [\int \frac{dq}{x_2}]^2\}x_2 , \quad x_2 = x_2, \quad x_3 = 0,$$

(26)

$$x_4 = \alpha x_1 , \quad x_5 = -x_1 + \alpha x_2 = \frac{1}{2}\{\alpha + [\int\frac{dq}{x_2}]^2\}x_2 ,$$

$$x_6 = x_2 , \quad x_7 = \{\int\frac{dq}{x_2}\} x_2.$$

The equations for q_x and q_t have the common factor $x_2(q)$ and a transformation of the type (12) yields the rather unexpected result that the pseudopotential,

$[y_8]$, so effectively used by Wahlquist and Estabrook [1] is in fact, to within the above transformation, the pseudopotential of the first kind. The Bäcklund transformation constructed by them is the most general possible, within the framework defined above. It should be stressed that the computational advantage of realizing the nonessential nature of the common factor x_2 is considerable. In particular, the computations necessary to obtain a new solution of KdV equation in terms of q, u, z, and p is greatly simplified.

It should further be noted that the $x_k(q)$ form a Lie algebra. In particular, after applying a transformation of the type (12), we note that

$$\tilde{x}_1 = -\frac{\alpha}{2} + \frac{1}{2}\tilde{q}^2, \quad \tilde{x}_2 = 1, \quad \tilde{x}_7 = \tilde{q}, \tag{27}$$

are the only linearly independent $\tilde{x}_j$ and it is easily seen that

$$\begin{aligned} &[\tilde{x}_1,\tilde{x}_2] = \tilde{x}_7 \qquad [\tilde{x}_1,\tilde{x}_7] = x_1 + \alpha x_2 \\ &[\tilde{x}_2,\tilde{x}_7] = -\tilde{x}_2 \end{aligned} \tag{28}$$

The Bäcklund transformation and hence linear problem for the KdV equation follow from a representation of this structure.

Although the calculation of general pseudopotentials of the nth-kind becomes complicated, the special case where F^k and G^k are linear in the pseudopotentials q^k, motivated by the method of inverse scattering, is technically feasible.

IV. LINEAR PROLONGATION STRUCTURES

It is clear by inspection that any linear first-order eigenvalue equation (together with its "time" evolution) used in the inverse method is a linear pseudopotential. The advantage of looking at linear problems this way is that it may be possible, particularly for a given dimension, say $m = 2$, to decide if there is any possible associated first-order linear eigenvalue problem for the given equation. If it can be established that there is no nonabelian two-dimensional prolongation structure, then there can be no two-dimensional first-order linear problem. Hence, consider the system

$$F = K(z^{\mu})q + L(z^{\mu}) \ , \quad G = M(z^{\mu})q + P(z^{\mu}) \tag{29}$$

where $q(x,y)$ is an m-dimensional vector, K, L, M, P are $m\times m$ matrices, and free indices are suppressed for convenience. Application of (5) produces the complete primitive variable dependence of the matrices K, L, M, P, leaving a set of commutation relations among a fundamental set of constant matrices (since all variable dependences have been either computed or assumed) whose existence is contingent upon the form of (2).

1. The Hirota Equation

An interesting example of the calculation of linear prolongation structures is found by considering the complex valued function $\phi(x,t)$, satisfying the "Hirota equation" [3] given by

$$\phi_t = -\ 3\alpha\phi\bar{\phi}\phi_x - \beta\phi_{xxx} + i\gamma\phi_{xx} + i\epsilon\phi^2\bar{\phi}, \tag{30}$$

and the associated primitive set z^{μ} of the form

$$z^{\mu} \equiv \{\phi,\bar{\phi},\phi_x,\bar{\phi}_x,\phi_{xx},\bar{\phi}_{xx}\}. \tag{31}$$

For convenience, we specialize the system (29) to the forms

$$F \equiv (\phi A + \bar{\phi}B + C)q \ , \quad G \equiv Dq, \tag{32}$$

where A, B, C, D are $m\times m$ matrices to be determined from (5) subject to (30). For technical reasons, we make the ansatz

$$[B,A] = a_1C + a_2I, \tag{33}$$

for some constants a_1, a_2 and where I is the $m\times m$ identity matrix. In the case $m = 2$, the ansatz in (33) is equivalent to the assumption that C have distinct eigenvalues (as a matrix). From (5), it then follows that A, B, C are constant matrices satisfying the commutation relations

(34) $$\beta a_1[B,C] = \alpha B,\ \beta a_1[A,C] = -\alpha A,$$

subject to (30), D is the matrix

$$\begin{aligned} D = {} & \{-\beta\bar{\phi}_{xx} + \mu\bar{\phi}_x - \alpha\bar{\phi}\phi^2 + a_3\bar{\phi}\}B \\ & + \{-\beta\phi_{xx} - \mu\phi_x - \alpha\phi\bar{\phi}^2 + a_3\phi\}A \\ & + \{\beta\phi\bar{\phi}_x - \beta\bar{\phi}\phi_x - \mu\phi\bar{\phi}\}\ (a_1C + a_2I) + a_3C, \end{aligned} \tag{35}$$

with the definitions

(36) $$a_3 \equiv \frac{\alpha}{\beta a_1}\mu\ ,\quad \mu \equiv \frac{\alpha}{a_1} - i\epsilon\ ,$$

resulting in functions F and G satisfying (5) provided that

(37) $$\beta\epsilon - \alpha\gamma = 0.$$

Although the necessity of (37) for existence of linear prolongation structures associated with (30) has not been established in general, it is very interesting that (37) has been shown to be a necessary and sufficient condition for the existence of n-soliton solutions. It is also important to note that existence of solutions to the commutation relations (34) subject to (33) is guaranteed for all values of m by standard results in matrix theory, provided that the matrix elements in C and the parameter a_1 are suitably restricted. Finally, it should be noted that in this case as well bona fide Lie algebra structure is present in (33) and (34). In this case the structure was forced by the ansatz (33) although a more careful study of prolongation structures may well lead to (33) as a necessary condition.

2. The Burgers-Modified Korteweg-deVries Equation

Another interesting example of an equation exhibiting a linear prolongation structure is given by the Burgers-Modified-Korteweg-deVries (B-MKdV) equation

(38) $$u_t = u_{xxx} + au_{xx} - bu^2u_x,\ b > 0,$$

with the primitive set z^μ taken in the form

(39) $$z^\mu \equiv \{u, z \equiv u_x, p \equiv u_{xx}\} .$$

Substitution of (29) into (5) with $L = P = 0$ then gives the results

(40) $$\begin{aligned} K &= uB + C \\ M &= \{p + az - \frac{bu^3}{3}\}B + (z + au)[C,B] \\ &\quad + \frac{u^2}{2}[B,[C,B]] + u[C,[C,B]] + D, \end{aligned}$$

where B, C, D are constant mxm matrices satisfying the commutation relations

(41) $$\begin{aligned} &\frac{3}{2}[B,[B,[C,B]]] = b[C,B], \quad [C,D] = 0, \\ &a[B,[C,B]] + \frac{3}{2}[B,[C,[C,B]]] = 0, \\ &[B,D] + a[C,[C,B]] + [C,[C,[C,B]]] = 0. \end{aligned}$$

Particular solutions to the commutation relations (41) in any dimension m are easily obtained by means of the ansatz

(42) $$[C,B] \equiv \rho C + \epsilon B, \quad D \equiv \gamma C,$$

with the definitions

(43) $$\rho \equiv (\frac{2}{3}b)^{1/2}, \quad \epsilon \equiv -\frac{2}{3}a, \quad \gamma \equiv -(a + \epsilon),$$

yielding the results

(44) $$\begin{aligned} K &= uB + C, \\ M &= \{p + (a + \epsilon)z - \frac{bu^3}{3} - \rho\frac{\epsilon u^2}{2} - \gamma u\}B \\ &\quad + \{\rho z + \rho(a + \epsilon)u - \rho^2\frac{u^2}{2} + \gamma\}C . \end{aligned}$$

The ansatz (42) appears to admit nontrivial solutions in any dimension m. It should be noted that the related Burgers-KdV equation $u_t = u_{xxx} + au_{xx} - buu_x$ possesses no nontrivial prolongation structure in two dimensions.

3. Associated Eigenvalue Problem

It is not yet clear if the existence of a linear prolongation structure implies the existence of an eigenvalue problem and associated evolution operator that is at all useful for solving the original equation. The reason for this is that first-order eigenvalue problems are in some cases obviously degenerate with respect to the inverse method of solution.

For example, consider

(45) $$\psi_x + u\psi = \lambda\psi,$$

(46) $$\psi_t = \int^x K(u)\,dx\,\psi,$$

where ψ is a scalar function and

(47) $$u_t = K(u).$$

The eigenvalues of (45) are preserved if Ψ evolves according to (46) for <u>any</u> equation (47).

A higher-dimensional "universal example" is provided by

(48) $$\hat{q}_x + C(z^\mu)\hat{q} = \lambda D\hat{q}.$$

(49) $$\hat{q}_t = B(z^\mu)\hat{q},$$

The eigenvalues of (45) are preserved if the n-component vector $\hat{q}$ evolves according to (46) provided (47) is satisfied. An example of this is when B commutes with D and

(50) $$C = uA\ ,\ B = -\int^x K(u)\,dx\ A$$

where A is an arbitrary matrix.

It is clear that satisfying a compatibility condition, i.e. the "cross-differentiation" condition of [4], <u>alone</u> is not sufficient to insure an eigenvalue problem is "good" from the point of view of the inverse method. Looked at as prolongation structures (45) and (46) are trivial by virtue of (12) and (48) and (49) are abelian. It does not seem possible to construct universal nonabelian linear pseudopotentials, i.e. to find a given functional form for linear pseudopotentials such that the associated prolongation structure is nonabelian and the pseudopotential is defined for any partial differential equation. The linear pseudopotential given for the B-MKdV equation is not a specialized universal example. The pseudopotential depends critically on the full structure of the equation.

Given a nonabelian linear pseudopotential, can a linear eigenvalue problem always be found? Certainly if

$$q_x = A(z^n)q, \tag{51}$$

$$q_t = B(z^n)q, \tag{52}$$

and

$$A_t - B_x + [A,B] = 0, \tag{53}$$

then

$$\tilde{q}_x - A(z^n)\tilde{q} = \lambda C\tilde{q}, \tag{54}$$

$$\tilde{q}_t = B(z^n)\tilde{q}, \tag{55}$$

(where all λ dependence is explicit) are <u>bona fide</u> eigenvalue and time evolution equations, provided $[C,B] = 0$. This can always be satisfied if C is the identity. Unfortunately the "empirical" evidence is that C must be traceless ([4], [5]). However, no general discussion of this point has as yet been given.

It can also be demanded that the eigenvalue problems arise in a more "organic" fashion from the pseudopotentials. For example, from the discussion of the Hirota equation, in (32), it could be demanded that $C = \lambda C'$ and that neither <u>A nor B depend on λ</u>. This is an added restriction on the prolongation structure. In the case of the Hirota equation this restriction can be satisfied. In two dimensions a linear problem and associated time evolution equation are obtained for the Hirota equation. The same problem is implicit in the general results of [4].

V. CONCLUDING REMARKS

The realization that the existence of a linear prolongation structure is necessary for the existence of an associated first-order linear problem coupled with the observation (as indicated in [1] and this work) that pseudopotentials exist rather infrequently clearly show how limited is the scope of the inverse method. Of equal importance, it raises the question of what are the sufficient conditions for a linear problem to be useful within this formalism, a question which, to our knowledge, has not been treated thus far. The B-MKdV may or may not be solvable by the inverse method but it seems to offer an ideal test case for conjectures in this area.

As was indicated earlier, abelian prolongation structures yield potentials and therefore "classical" conservation laws associated with a given equation. These conservation laws involve at most the highest derivative of the solution of the given equation that was present in the original primitive set of variables. For the KdV equation this set is u, $z = u_x$, $p = u_{xx}$. To obtain higher conservation laws it is necessary to start with a larger primitive set, adding say $r = u_{xxx}$ to the original. It is not yet clear if pseudopotentials of the first kind exist for the KdV equation based on this larger primitive set. That is, do there exist F and G such that

(56a) $$q_x = F(u,z,p,r,q)$$

(56b) $$q_t = G(u,z,p,r,q)$$

are nontrivial and integrable where u is subject to the KdV equation. Analogous questions can be asked for other equations. If F and G do exist and are computable, it would be of considerable interest to check if a Bäcklund transformation based on (56) can be obtained; a Bäcklund transformation which might have more than one free parameter. The question of associated eigenvalue problems might also be considered in this context.

ACKNOWLEDGMENT

We are indebted to F. Estabrook and H. Wahlquist for several very useful discussions and for reading an earlier version of the manuscript.

REFERENCES

[1] H.D. WAHLQUIST AND F.B. ESTABROOK, Prolongation structures of nonlinear evolution equations, J. Mathematical Phys. 16 (1975), 1-7.

[2] D.W. MCLAUGHLIN AND A.C. SCOTT, A restricted Bäcklund transformation, J. Mathematical Phys. 14 (1973), 1817-1828.

[3] R. HIROTA Exact envelope-soliton solutions of a nonlinear wave equation, J. Mathematical Phys. 14 (1973), 805-809.

[4] M.J. ABLOWITZ, D.J. KAUP, A.C. NEWELL, AND H. SEGUR, The inverse scattering transform-Fourier analysis for nonlinear problems, Studies in Appl. Math. 53 (1974), 249-315.

[5] V.E. ZAKHAROV AND S.V. MANAKOV, Resonant interaction of wave packets in nonlinear media, Soviet Physics JETP Lett. 18 (1973), 243-245.

VARIATIONAL PROBLEMS AND BÄCKLUND TRANSFORMATIONS ASSOCIATED WITH THE SINE-GORDON AND KORTEWEG-DEVRIES EQUATIONS AND THEIR EXTENSIONS*

Hanno Rund

Department of Mathematics
University of Arizona
Tucson, Arizona 85721

I. INTRODUCTION

There does not appear to be an authoritative, generally accepted definition of the concept of a Bäcklund transformation, and accordingly a somewhat naive point of view will be adopted in the present approach. We shall consider a pair of partial differential equations, expressed briefly in the form

$$E(x) = 0 , \tag{1.1}$$

$$D(y) = 0 , \tag{1.2}$$

in which x, y denote the unknown functions, while E and D represent differential operators in m independent variables. A system consisting of one or more relations involving x, y and their derivatives (up to any required order) will be called a Bäcklund transformation if these relations ensure that y satisfies (1.2) whenever x satisfies (1.1), and conversely.

There are, of course, many different ways in which this may be achieved, and accordingly our point of departure is a fairly obvious connection between transformations of this kind and the calculus of variations. Thus it is assumed initially that (1.1) is the Euler-Lagrange equation of a variational principle with given Lagrangian $L(x)$. If there exist relations between the functions x and y and their derivatives which are such as to imply that the difference $L(y) - L(x)$ is a divergence, then $E(y) = 0$ whenever (1.1) is satisfied, and conversely. Thus the relations in question possess the desired property for the pair of equations $E(x) = 0$, $E(y) = 0$, and are therefore called <u>variational</u> Bäcklund transformations.

*This research was supported in part by NSF GP-43070.

The above criterion is applied to a class of second-order partial differential equations in m independent variables which contain the sine-Gordon equation as a special case when $m = 2$. The method not only yields the variational Bäcklund transformations almost effortlessly, but also clearly indicates the circumstances under which they exist. However, when we turn to the next case, namely that of the Korteweg-deVries equation together with its modifications, it is to be observed that the latter are merely closely related to--but not identical with--the Euler-Lagrange equations of a variational problem, and accordingly certain adaptations of the technique are required. This in turn suggests that in this case it may be preferable to consider the so-called simple Bäcklund transformations, which are defined by the property that they guarantee that the difference $E(y) - E(x)$ vanishes, and the example of one of the modified Korteweg-deVries equations shows that a simple Bäcklund transformation need not be a variational one. Nevertheless, a criterion for the existence of simple Bäcklund transformations for equations of this type may be established.

An inspection of the variational Bäcklund transformations obtained for the generalized sine-Gordon type equations reveals that these are not only simple, but in fact imply that $E(x)$ and $E(y)$ vanish separately. Bäcklund transformations possessing this property are said to be strong, and appear to be particularly useful when one is concerned with pairs of partial differential equations which cannot be associated with variational principles. This is illustrated for the case when (1.1) and (1.2) are the diffusion and Burgers equations, respectively. For this example a simple Bäcklund transformation can be written down immediately in terms of an entirely arbitrary function: by demanding that the latter be such as to ensure that the transformation is also a strong one, two sets of transformations are obtained, one of which is the well-known Hopf-Cole tranformation. A very similar procedure may also be applied to a discussion of the transformation found by Miura, which relates the Korteweg-deVries equation to one of its modifications.

Needless to say, many open questions remain. It is fairly obvious that one cannot expect all known Bäcklund transformations to be amenable to a classification and corresponding treatment of this kind; for instance, the Bäcklund transformation which appears in the theory of 2-dimensional gas dynamics does not fall within the aforementioned categories, as is immediately evident from the fact that these transformations also involve a change of the independent variables. We hope to deal with extensions of this kind presently.

II. VARIATIONAL THEORY

We shall consider an m-fold integral variational problem whose m independent variables are denoted by t^α, the n dependent functions being represented by x^j. (Lower case Greek and Latin indices range from 1 to m and from 1 to n, respectively; the summation convention applies to both sets.) In the configuration space, R^{m+n}, of the variables (t^α, x^j) a system of n equations of the form $x^j = x^j(t^\alpha)$ represents an m-dimensional subspace C_m, whose tangent plane at each point $P(t^\beta, x^j)$ is spanned by the m vectors whose components in R^{m+n} are $(\delta^\beta_\alpha, \dot{x}^j_\alpha)$, where $\dot{x}^j_\alpha = \partial x^j/\partial t^\alpha$ (it being assumed henceforth that the functions $x^j(t^\alpha)$ are continuously differentiable up to any required order). For a given Lagrange function $L(t^\alpha, x^j, \dot{x}^j_\alpha)$, assumed to be of class C^2 in all its arguments and denoted henceforth simply by $L(x)$, we define the m-fold fundamental integral

$$I(x) \equiv \int_G L(x)d(t) , \tag{2.1}$$

where G denotes a bounded, simply-connected region in the domain R^m of the independent variables t^α and $d(t)$ represents the m-dimensional volume element of G. In general the value of $I(x)$ depends on the choice of C_m. In order that C_m afford an extreme value to $I(x)$ it is necessary that it represents a solution of the Euler-Lagrange equations

$$E_j(x) = 0, \quad j = 1,\dots,n , \tag{2.2}$$

where

(2.3)
$$E_j(x) \equiv \frac{d}{dt^\alpha}\left(\frac{\partial L}{\partial \dot{x}^j_\alpha}\right) - \frac{\partial L}{\partial x^j} ,$$

in which the operator d/dt^α is defined as

(2.4)
$$\frac{d}{dt^\alpha} \equiv \frac{\partial}{\partial t^\alpha} + \dot{x}^j_\alpha \frac{\partial}{\partial x^j} + \ddot{x}^j_{\alpha\beta} \frac{\partial}{\partial \dot{x}^j_\beta} .$$

Now let us assume that, relative to some other set of n dependent functions $y^j(t^\alpha)$, there exist p relations of the form

(2.5)
$$F^A(t^\alpha, x^j, y^j, \dot{x}^j_\alpha, \dot{y}^j_\alpha) = 0, \quad A = 1,\ldots,p ,$$

which are such as to entail that the difference $L(y) - L(x)$ is a divergence, i.e.,

(2.6)
$$L(y) - L(x) = \frac{d\Phi^\alpha}{dt^\alpha} ,$$

where $\Phi^\alpha = \Phi^\alpha(t^\beta, x^j, y^j, \dot{x}^j_\beta, \dot{y}^j_\beta)$ denotes a suitable set of m functions satisfying the skew-symmetry conditions

(2.7)
$$\frac{\partial \Phi^\alpha}{\partial \dot{x}^j_\beta} = -\frac{\partial \Phi^\beta}{\partial \dot{x}^j_\alpha} , \quad \frac{\partial \Phi^\alpha}{\partial \dot{y}^j_\beta} = -\frac{\partial \Phi^\beta}{\partial \dot{y}^j_\alpha} ,$$

which ensure that the right side of (2.6) is independent of the second-order derivatives $\ddot{x}^j_{\alpha\beta}$, $\ddot{y}^j_{\alpha\beta}$. Subject to this assumption it is now asserted that <u>the equations</u> (2.2) <u>imply that</u>

(2.8)
$$E_j(y) = 0 ,$$

<u>and conversely</u>. In this sense, therefore, the relations (2.5) play the role of a Bäcklund transformation, and will henceforth be called <u>variational Bäcklund transformations</u>.

In order to prove this assertion, we apply the divergence theorem to (2.6), which yields

(2.9)
$$\int_G L(y)d(t) - \int_G L(x)d(t) = \int_{\partial G} n_\alpha \Phi^\alpha d(\tau) ,$$

where n_α denotes the unit outward normal in R^m to the boundary ∂G of G and $d(\tau)$ represents the $(m-1)$-dimensional volume element of ∂G, the latter being supposed to be sufficiently smooth. The integral on the right side of (2.9) clearly depends solely on prescribed values of $x, y, \dot{x}^j_\alpha, \dot{y}^j_\alpha$ on ∂G (which must, of course, be consistent with (2.5)). Keeping these values fixed, we proceed as usual to apply a variation to the functions x^j of the form

$$\bar{x}^j(t^\alpha) = x^j(t^\alpha) + \varepsilon\xi^j(t^\alpha) ,$$

where the class C^1 functions $\xi^j(t^\alpha)$ are arbitrary except for the conditions $\xi^j(t^\alpha) = 0$ and $\dot{\xi}^j_\beta(t^\alpha) = 0$ on ∂G. Because of (2.5), this will imply some variation

$$\bar{y}^j(t^\alpha) = y^j(t^\alpha) + \varepsilon\eta^j(t^\alpha)$$

of the functions y^j, with $\eta^j(t^\alpha) = 0$, $\dot{\eta}^j_\beta(t^\alpha) = 0$ on ∂G (which is required for consistency with (2.5)). From the general theory of the first variation of multiple integrals in the calculus of variations [6, pp. 210-217] it then follows that

$$\int_G E_j(x)\xi^j d(t) = \int_G E_j(y)\eta^j d(t) .$$

Now, if y satisfies (2.8), the integral on the left vanishes, and for a variation with $\xi^2 = \xi^3 = \ldots = \xi^n = 0$, we thus have

$$\int_G E_1(x)\xi^1 d(t) = 0 . \tag{2.10}$$

Let us suppose, for the moment, that $E_1(x) > 0$ at some point $P(t_0)$ in the interior of G. Then there exists a number ρ such that $E_1(x) > 0$ in the ball $B(P,\rho)$, where $B(P,\rho) \subset G$. The function $\xi^1(t^\beta)$ is now defined on G by putting

$$\xi^1(t^\beta) = [\rho^2 - \sum_{\alpha=1}^{m} (t^\alpha - t_0^\alpha)^2]^4$$

for $t^\alpha \in B(P,\rho)$, and $\xi^1(t^\beta) = 0$ otherwise. The resulting variation satisfies all previous requirements on ∂G, but renders the integral (2.10) positive, thus yielding a contradiction between (2.10) and the assumption that $E_1(x) > 0$ at P. Similarly, the assumption that $E_1(x) < 0$ at P is also contradictory, and hence we conclude that $E_1(x) = 0$ on G. In the same manner, it is shown that $E_2(x) = \ldots = E_n(x) = 0$. Thus (2.8) implies (2.2); the converse is established by simply interchanging the roles of x and y in the above argument.

For future reference we observe that the above analysis is easily extended to second-order variational problems, that is, to problems for which $L(x) = L(t^\alpha, x^j, \dot{x}^j_\alpha, \ddot{x}^j_{\alpha\beta})$. In this case the corresponding Euler-Lagrange equations (2.2) are generally of the fourth order, with $E_j(x)$ given by

$$(2.11) \qquad E_j(x) = \frac{d}{dt^\alpha}\left[\frac{\partial L}{\partial \dot{x}^j} - \frac{d}{dt^\beta}\left(\frac{\partial L}{\partial \ddot{x}^j_{\alpha\beta}}\right)\right] - \frac{\partial L}{\partial x^j} \ .$$

However, under these circumstances the functions F^A and Φ^α which occur in (2.5) and (2.6) may also depend on the second-order derivatives of x^j and y^j.

III. EQUATIONS OF THE SINE-GORDON TYPE

As a first illustration of a variational Bäcklund transformation we shall consider a single partial differential equation of the form

$$(3.1) \qquad E(x) \equiv \Pi\Lambda(x) + f(x) = 0 \ ,$$

where Λ, Π are differential operators defined by

$$(3.2) \qquad \Lambda \equiv a^\alpha \frac{\partial}{\partial t^\alpha} \ , \quad \Pi \equiv b^\beta \frac{\partial}{\partial t^\beta} \ ,$$

in which a^α, b^β are $2m$ given constants. Clearly (3.1) is the Euler-Lagrange equation (2.2) of a variational principle (with $n = 1$, m arbitrary) whose Lagrangian is given by

$$(3.3) \qquad L(x) = \frac{1}{2}\Lambda(x)\Pi(x) - g(x) \ ,$$

where

$$g'(x) = f(x) \quad (3.4)$$

Initially it will be assumed that

$$g(x) = -\cos x , \quad (3.5)$$

since (3.1) will then contain the sine-Gordon equation as a special case ($m = 2$).

According to the prescription of the previous section, we now seek conditions under which there exist m functions Φ^α such that

$$L(y) - L(x) \equiv \frac{1}{2}\,[\Lambda(y)\Pi(y) - \Lambda(x)\Pi(x)] + \cos y - \cos x = \frac{d\Phi^\alpha}{dt^\alpha} . \quad (3.6)$$

To this end we put

$$u = \frac{1}{2}\,(y + x), \quad v = \frac{1}{2}\,(y - x) , \quad (3.7)$$

since it is easily verified that

$$\Lambda(y)\Pi(y) - \Lambda(x)\Pi(x) = 2[\Lambda(u)\Pi(v) + \Lambda(v)\Pi(u)] , \quad (3.8)$$

while

$$\cos y - \cos x = -2\sin u \sin v . \quad (3.9)$$

Thus

$$\begin{aligned} L(y) - L(x) &= \Lambda(u)\Pi(v) - \Lambda(v)\Pi(u) + 2\{\Lambda(v)\Pi(u) - \sin u \sin v\} \quad (3.10)\\ &= \Lambda[u\Pi(v)] - \Pi[u\Lambda(v)] + 2\{\Lambda(v)\Pi(u) - \sin u \sin v\}, \end{aligned}$$

where it is to be observed that all terms involving second-order derivatives have already been absorbed in the first two terms on the right side, which are automatically divergences. The required condition (3.6) is obviously satisfied provided that the remaining term in braces is also a divergence, that is, if there exist functions $\phi(u,v)$, $\psi(u,v)$ such that

$$\Lambda(v)\Pi(u) - \sin u \sin v = \Lambda[\phi(u,v)] + \Pi[\psi(u,v)] \tag{3.11}$$

$$= \frac{\partial\phi}{\partial u}\Lambda(u) + \frac{\partial\phi}{\partial v}\Lambda(v) + \frac{\partial\psi}{\partial u}\Pi(u) + \frac{\partial\psi}{\partial v}\Pi(v) .$$

However, since the left side involves no terms in $\Lambda(u)$, $\Pi(v)$, the coefficients of the latter on the right side should vanish, i.e., $\partial\phi/\partial u = 0$, $\partial\psi/\partial v = 0$, or $\phi = \phi(v)$, $\psi = \psi(u)$, so that (3.11) reduces to

$$\Lambda(v)\Pi(u) - \sin u \sin v = \phi'(v)\Lambda(v) + \psi'(u)\Pi(u) . \tag{3.12}$$

A trial solution suggests itself by inspection of the coefficients of $\Lambda(v)$ and $\Pi(u)$ on either side, which leads to two distinct possibilities, namely, <u>either</u>

$$\phi'(v) = \Pi(u) , \tag{3.13}$$

<u>or</u>

$$\psi'(u) = \Lambda(v) . \tag{3.14}$$

<u>Case I</u>: Let us suppose that (3.13) holds, where it should be noted already at this stage that (3.13) is of the form (2.5). When (3.13) is substituted in (3.12), the latter becomes

$$\sin u \sin v = -\psi'(u)\phi'(v) , \tag{3.15}$$

which allows for a separation of variables, giving

$$\frac{\sin u}{\psi'(u)} = -\frac{\phi'(v)}{\sin v} = a^{-1} ,$$

where a is an arbitrary nonzero constant. Thus

$$\phi(v) = a^{-1}\cos v, \quad \psi(u) = -a \cos u , \tag{3.16}$$

and (3.13) assumes the explicit form

$$(3.17) \qquad \Pi(u) = -a^{-1}\sin v \ .$$

Clearly the single relation (3.17) is sufficient to ensure that the condition (3.6) is satisfied; in fact, it is found by substitution of (3.17) in (3.10) that

$$(3.18) \qquad L(y) - L(x) = \Lambda[u\Pi(v) + 2a^{-1}\cos v] - \Pi[u\Lambda(v) + 2a\cos u] \ ,$$

from which it follows that the functions Φ^α which appear in (3.6) are given explicitly by

$$(3.19) \qquad \Phi^\alpha = (a^\alpha b^\beta - b^\alpha a^\beta)u\dot{v}_\beta + 2(a^{-1}a^\alpha\cos v - ab^\alpha\cos u) \ .$$

(Incidentally, we observe that the skew-symmetry condition (2.7) is automatically satisfied by (3.19).) We therefore conclude that (3.17) is a variational Bäcklund transformation.

Case II: Instead of (3.13), it is now assumed that (3.14) holds. When (3.14) is substituted in (3.12), the latter again reduces to (3.15), yielding the solutions (3.16). Thus (3.14) is equivalent to

$$(3.20) \qquad \Lambda(v) = a\sin u \ ,$$

which is obviously the second variational Bäcklund transformation (for which the associated functions Φ^α also assume the form (3.19)).

For the case when $m = 2$ we can, without loss of generality, assume that a^α, b^α are such that $\Lambda = \partial/\partial t^1$, $\Pi = \partial/\partial t^2$, so that (3.1) reduces to

$$(3.21) \qquad \ddot{x}_{12} + f(x) = 0 \ .$$

If it is also supposed that $f(x) = \sin x$, the relations (3.17) and (3.20) become

$$\tfrac{1}{2}(\dot{x}_2 + \dot{y}_2) = -a^{-1}\sin\{\tfrac{1}{2}(y-x)\}, \qquad \tfrac{1}{2}(\dot{y}_1 - \dot{x}_1) = a\sin\{\tfrac{1}{2}(y+x)\} \ ,$$

which represent, of course, the well-known Bäcklund transformations associated with the sine-Gordon equation [2]. This immediately raises the question as to whether (3.21) admits Bäcklund transformations for choices of $f(x)$ other than $f(x) = \sin x$. In fact, on retracing the steps (3.6) to (3.11) for an arbitrary function $g(x)$, with $g'(x) = f(x)$, it is found that $L(y) - L(x)$ cannot be cast into the form of a divergence by means of a variational Bäcklund transformation unless $g(x)$ is such as to satisfy a functional equation of the type

$$g(u+v) - g(u-v) = F(u)G(v) \ . \tag{3.22}$$

But this is a <u>generalized d'Alembert equation</u>, and one may therefore apply a powerful theorem [1,p. 176] in order to infer that there exist only three distinct solutions, namely

$$g(x) = \alpha \cos bx + \beta \sin bx + \gamma \ ,$$

$$g(x) = \alpha \cosh bx + \beta \sinh bx + \gamma,$$

$$g(x) = \alpha x^2 + \beta x + \gamma \ ,$$

where b, α, β, γ are arbitrary constants. From (3.4) it then follows that the equation (3.21) admits variational Bäcklund transformations if and only if $f(x)$ possesses one of the following three forms:

$$f(x) = A \cos bx + B \sin bx \ ,$$

$$f(x) = A \cosh bx + B \sinh bx \ , \tag{3.23}$$

$$f(x) = Ax + B \ ,$$

in which A and B are arbitrary constants. It should be remarked that these restrictions on equation (3.21) are also given by McLaughlin and Scott [4], whose approach, however, is based on an entirely different definition of Bäcklund transformation.

This particular complex of ideas suggests a more direct alternative approach. Returning to the case of arbitrary values of m, we note that, corresponding to the Lagrangian (3.3), we have by virtue of (3.1) and (3.7),

(3.24) $$E(y) - E(x) = 2\Pi\Lambda(v) + f(u+v) - f(u-v) \ .$$

A pair of relations of the type

(3.25) $$\Pi(u) = \lambda(v), \quad \Lambda(v) = \theta(u) \ ,$$

is said to represent a simple Bäcklund transformation if the functions $\lambda(v)$, $\theta(u)$ are such as to ensure that the application of (3.25) to (3.24) results in the condition

(3.26) $$E(y) - E(x) = 0 \ .$$

Since (3.25) implies that

(3.27) $$\Pi\Lambda(v) = \theta'(u)\Pi(u) = \theta'(u)\lambda(v) \ ,$$

it follows that (3.26) is satisfied as a consequence of (3.25) provided that

(3.28) $$f(u+v) - f(u-v) = -2\theta'(u)\lambda(v) \ .$$

However, this is again a generalized d'Alembert equation, and the aforementioned theorem from the theory of functional equations leads to the conclusion that $f(x)$ can possess only one of the three forms displayed in (3.23).

Let us consider the first of these, substituting the given form of $f(x)$ into (3.28), obtaining

$$\sin bv[A \cos bu - B \sin bu] = -\theta'(u)\lambda(v) \ ,$$

from which it is immediately inferred that

(3.29) $$\lambda(v) = -a^{-1}\sin bv, \quad \theta(u) = ab^{-1}(A \sin bu + B \cos bu) \ .$$

Thus the simple Bäcklund transformation (3.25) is completely determined. Moreover, it is obvious that the same procedure can be applied with equal facility to the other two cases listed in (3.23). Our conclusions may be summarized in the following

Theorem: The partial differential equation (3.1) admits a simple Bäcklund transformation if and only if the function $f(x)$ possesses one of the three forms displayed in (3.23), in which case the corresponding Bäcklund transformations are respectively given by

$$\Pi(u) = -a^{-1}\sin bv,\ \Lambda(v) = ab^{-1}[A \sin bu + B \cos bu]\ ,$$

$$\Pi(u) = -a^{-1}[A \cosh bv + B \sinh bv],\ \Lambda(v) = ab^{-1}\cosh bu\ , \tag{3.30}$$

$$\Pi(u) = a^{-1}v,\ \Lambda(v) = -aAu - aB\ .$$

It should be observed that these simple transformations are also variational Bäcklund transformations, as is easily verified by the construction of suitable Lagrangians for each of the three cases under consideration. In general, however, the concepts of simple and variational Bäcklund transformations do not coincide. Moreover, the example studied in this section exhibits another special feature: the transformations (3.30) do not merely guarantee the validity of the condition (3.26), but each pair in (3.30) implies the much stronger consequence that $E(x)$ and $E(y)$ vanish separately, as follows directly from an evaluation of $\Pi\Lambda(u+v)$ and $\Pi\Lambda(u-v)$ in each case. Bäcklund transformations possessing this property will be called strong; we shall return to this point presently.

IV. THE KORTEWEG-DEVRIES AND RELATED EQUATIONS

We shall first consider a variational problem for which $m = 2$, whose Lagrangian is assumed to be of the second order, namely

$$L(x) = \frac{1}{2}\dot{x}_1\dot{x}_2 - \frac{1}{2}(\ddot{x}_{11})^2 + \frac{\alpha}{n(n+1)}(\dot{x}_1)^{n+1}\ , \tag{4.1}$$

where α, n are given constants. The general Euler-Lagrange expression (2.11) with (4.1) is given by

(4.2) $$E(x) = \frac{d}{dt^1}\,[\dot{x}_2 + \frac{\alpha}{n}(\dot{x}_1)^n + \dddot{x}_{111}]\ .$$

Thus, for the Lagrangian (4.1), we have a most unusual situation in that the Euler-Lagrange expression is a perfect t^1-derivative. Written out in full, the expression (4.2) becomes

(4.3) $$E(x) = \ddot{x}_{12} + \alpha(\dot{x}_1)^{n-1}\ddot{x}_{11} + \ddddot{x}_{1111}\ ,$$

which suggests that, for any function ϕ, we should define an operator K by writing

(4.4) $$K(\phi) = \dot{\phi}_2 + \alpha(\phi)^{n-1}\dot{\phi}_1 + \dddot{\phi}_{111}\ ,$$

so that (4.3) can be expressed in the form

(4.5) $$E(x) = K(\dot{x}_1)\ .$$

It should be noted that <u>the partial differential equation</u>

(4.6) $$K(\phi) = 0$$

<u>is identical with the Korteweg-deVries equation when</u> $n = 2$, <u>and with the modified Korteweg-deVries equation when</u> $n = 3$.

Let x be a solution of the Euler-Lagrange equation

(4.7) $$E(x) = 0\ .$$

It then follows from (4.5) that $\phi \equiv \dot{x}_1$ is a solution (4.6). Conversely, if $\phi = \phi(t^1,t^2)$ is a solution of (4.6), then the function

(4.8) $$x(t^1,t^2) = \int_{t_0}^{t^1} \phi(t,t^2)\,dt$$

is a solution of (4.7), and from (4.2) it is thus inferred that this function satisfies the condition

(4.9) $$E(x) = \psi(t^2)\ ,$$

where $\psi(t^2)$ is an arbitrary function of t^2 only, and

$$(4.10) \qquad E(x) \equiv \dot{x}_2 + \frac{\alpha}{n}(\dot{x}_1)^n + \dddot{x}_{111} .$$

Moreover, from (4.2) it is evident that <u>any solution</u> $x(t^1,t^2)$ <u>of the partial differential equation</u>

$$(4.11) \qquad E(x) = 0$$

<u>satisfies the Euler-Lagrange equation</u> (4.7) <u>and therefore generates the solution</u> $\phi = \dot{x}_1$ <u>of</u> (4.6).

According to the general theory we now seek one or more relations between the functions x, y which are such as to ensure that the expression $L(y) - L(x)$ is a divergence. We therefore begin by evaluating the difference $L(y) - L(x)$. In terms of the notation (3.7) we find that

$$(4.12) \qquad L(y) - L(x) = \frac{d}{dt^1}[v\dot{u}_2] - \frac{d}{dt^2}[v\dot{u}_1] + 2\Phi ,$$

where

$$(4.13) \quad \Phi \equiv \dot{u}_1\dot{v}_2 - \ddot{u}_{11}\ddot{v}_{11} + \frac{\alpha}{2n(n+1)}[(\dot{u}_1 + \dot{v}_1)^{n+1} - (\dot{u}_1 - \dot{v}_1)^{n+1}] .$$

Alternatively, we can also express (1.19) as

$$(4.14) \qquad L(y) - L(x) = \frac{d}{dt^2}[v\dot{u}_1] - \frac{d}{dt^1}[v\dot{u}_2] + 2\Psi ,$$

where

$$(4.15) \quad \Psi \equiv \dot{u}_2\dot{v}_1 - \ddot{u}_{11}\ddot{v}_{11} + \frac{\alpha}{2n(n+1)}[(\dot{u}_1 + \dot{v}_1)^{n+1} - (\dot{u}_1 - \dot{v}_1)^{n+1}] .$$

Our program now requires that we seek relations between the functions x, y and their derivatives which ensure that either Φ or Ψ be a divergence. Once this has been achieved, and if x is a solution of (4.11)--which implies that it is a solution of (4.7)--it follows that y is a solution of $E(y) = 0$. Thus the condition that either Φ or Ψ is a divergence is not quite sufficient to ensure that $E(y) = 0$ (which is what is actually required), and therefore one cannot proceed in exactly the same way as for the sine-Gordon equation (the latter being a true Euler-Lagrange equation). We shall now consider separately various cases which correspond to different values of n in (4.1).

1. The Korteweg-deVries Equation: n = 2

In this case (4.13) becomes

$$\Phi = \frac{\alpha}{6} \frac{d}{dt^1} [v\dot{v}_1^2] + \Phi^* , \tag{4.16}$$

where

$$\Phi^* \equiv (\dot{u}_1\dot{v}_2 - \ddot{u}_{11}\ddot{v}_{11}) + \frac{\alpha}{2} \dot{u}_1^2\dot{v}_1 - \frac{\alpha}{3} v\dot{v}_1\ddot{v}_{11} . \tag{4.17}$$

We now seek a function $\lambda(v)$ which is such as to ensure that the relation

$$\dot{u}_1 = \lambda(v) \tag{4.18}$$

renders Φ^* a divergence, noting that for any differentiable function $\lambda(v)$ we have

$$\lambda(v)\dot{v}_2 = \frac{d}{dt^2} [\mu(v)], \quad \lambda^2(v)\dot{v}_1 = \frac{d}{dt^1} [\nu(v)] ,$$

where

$$\mu'(v) = \lambda(v), \; \nu'(v) = \lambda^2(v) . \tag{4.19}$$

Assuming, then, that (4.18) is valid, we substitute the latter in (4.17), obtaining

$$\Phi^* = \frac{d}{dt^2} [\mu(v)] + \frac{\alpha}{2} \frac{d}{dt^1} [\nu(v)] - \dot{v}_1\ddot{v}_{11}[\lambda'(v) + \frac{\alpha}{3} v] . \tag{4.20}$$

This is a divergence as required, provided that we choose $\lambda(v)$ such that

$$\lambda'(v) + \frac{\alpha}{3} v = b , \tag{4.21}$$

where b is an arbitrary constant, for under these circumstances we have

(4.22) $$\Phi^* = \frac{d}{dt^2}\,[\mu(v)] + \frac{d}{dt^1}\,[\tfrac{\alpha}{2}\,\nu(v) - \tfrac{b}{2}\,\dot{v}_1^2]\ .$$

From (4.21) it follows that

(4.23) $$\lambda(v) = a + bv - \frac{\alpha}{6}\,v^2\ ,$$

and (4.18) yields the *first component of the required variational Bäcklund transformation, namely*

(4.24) $$\dot{u}_1 = a + bv - \frac{\alpha}{6}\,v^2\ .$$

The second component can be found in various ways; for instance, we could seek a condition on $\dot{u}_2$ which renders the function Ψ as defined by (4.15) with $n = 2$ a divergence. However, since this method will not guarantee that $E(y) = 0$, we now proceed instead to examine the difference

(4.25) $$E(y) - E(x) \equiv 2\chi\ .$$

From (4.10), with $n = 2$, we find that

(4.26) $$\chi = \dddot{v}_{111} + \dot{v}_2 + \alpha\dot{u}_1\dot{v}_1\ .$$

Therefore, since

$$\lambda'(v)\dddot{v}_{111} = \frac{d}{dt^1}\,[\lambda'(v)\ddot{v}_{11} - \tfrac{1}{2}\,\lambda''(v)\dot{v}_1^2] + \tfrac{1}{2}\,\lambda'''(v)\dot{v}_1^3$$

identically, we see that

(4.27) $$\lambda'(v)\chi = \frac{d}{dt^2}[\lambda(v)] + \frac{d}{dt^1}[\tfrac{\alpha}{2}\,\lambda^2(v) + \lambda'(v)\ddot{v}_{11} - \tfrac{1}{2}\,\lambda''(v)\dot{v}_1^2] + \tfrac{1}{2}\,\lambda'''(v)\dot{v}_1^3\ .$$

But from (4.23) it follows that $\lambda'''(v) = 0$. Thus the condition $\chi = 0$ is satisfied, provided that

(4.28) $$\frac{d}{dt^1}\,[\tfrac{\alpha}{2}\,\lambda^2(v) + \lambda'(v)\ddot{v}_{11} - \tfrac{1}{2}\lambda''(v)\dot{v}_1^2] + \frac{d}{dt^2}\,[\lambda(v)] = 0\ .$$

This, however, implies that there exists a function U such that

$$(4.29)\qquad \dot{U}_1 = \lambda(v),\quad \dot{U}_2 = -\frac{\alpha}{2}\lambda^2(v) - \lambda'(v)\ddot{v}_{11} + \frac{1}{2}\lambda''(v)\dot{v}_1^2 .$$

On comparing this with (4.18), we infer that $U = u + \phi(t^2)$, where $\phi(t^2)$ is an arbitrary function of t^2 (which we take to be zero). Thus, with the aid of (4.24) we may write the second member of (4.29) as

$$(4.30)\qquad \dot{u}_2 = -b\ddot{v}_{11} + \frac{\alpha}{6}[2v\ddot{v}_{11} - \dot{v}_1^2 - 3\dot{u}_1^2] .$$

This is the second component of the required Bäcklund transformation. When $b = 0$, $\alpha = 6$, the system (4.24), (4.30) reduces to the Bäcklund transformation given by Wahlquist and Estabrook [7].

2. The Modified Korteweg-deVries Equation: $n = 3$

In this case (4.13) assumes the form

$$(4.31)\qquad \Phi = \frac{\alpha}{3}\frac{d}{dt^1}[v\dot{u}_1\dot{v}_1^2] + \Phi^* ,$$

where

$$(4.32)\qquad \Phi^* \equiv (\dot{u}_1\dot{v}_2 - \ddot{u}_{11}\ddot{v}_{11}) + \frac{\alpha}{3}\dot{u}_1^3\dot{v}_1 - \frac{\alpha}{3}v\dot{v}_1^2\ddot{u}_{11} - \frac{2\alpha}{3}v\dot{u}_1\dot{v}_1\ddot{v}_{11} .$$

Again it is supposed that Φ^* can be reduced to a divergence by means of a relation of the type (4.18), and when this is substituted in (3.3) we obtain

$$(4.33)\quad \Phi^* = \lambda(v)\dot{v}_2 + \frac{\alpha}{3}\lambda^3(v)\dot{v}_1 - \dot{v}_1\ddot{v}_{11}[\lambda'(v) + \frac{2\alpha}{3}v\lambda(v)] - \frac{\alpha}{3}v\lambda'(v)\dot{v}_1^3 .$$

This can be written in the form

$$(4.34)\qquad \Phi^* = \frac{d}{dt^2}[\nu(v)] + \frac{d}{dt^1}[\frac{\alpha}{12}\lambda^4(v) + \mu(v)\dot{v}_1^2] + \Phi^{**} ,$$

where

$$(4.35)\qquad \Phi^{**} \equiv -\dot{v}_1\ddot{v}_{11}[\lambda'(v) + \frac{2\alpha}{3}v\lambda(v) + 2\mu(v)] .$$

with

$$(4.36)\qquad \nu'(v) = \lambda(v) ,\quad 3\mu'(v) = -\alpha v\lambda'(v) .$$

Thus Φ^* is a divergence, provided that we choose $\lambda(v)$ such that

(4.37) $$\lambda'(v) + \frac{2\alpha}{3} v\lambda(v) + 2\mu(v) = b ,$$

where b is an arbitrary constant. We differentiate this with respect to v, after which we substitute from (4.36), obtaining

(4.38) $$\lambda''(v) + \frac{2\alpha}{3}\lambda(v) = 0 .$$

<u>Thus the first component of the variational Bäcklund transformation becomes</u>

(4.39a) $$\dot{u}_1 = \lambda(v) = a \sin \beta v, \quad \underline{\text{with}} \quad \beta = \sqrt{\frac{2\alpha}{3}} , \quad \text{if} \quad \alpha > 0$$

<u>and</u>

(4.39b) $$\dot{u}_1 = \lambda(v) = a \sinh \beta v, \quad \underline{\text{with}} \quad \beta = \sqrt{-\frac{2\alpha}{3}} , \quad \text{if} \quad \alpha < 0 .$$

In order to obtain the second component, we now evaluate the expression (4.25). From (4.10) it follows that, for $n = 3$,

(4.40) $$\chi = \dddot{v}_{111} + \dot{v}_2 + \frac{\alpha}{3} \dot{v}_1^3 + \alpha \dot{v}_1 \dot{u}_1^2$$

so that

(4.41) $$\lambda'(v)\chi = \frac{d}{dt^2} [\lambda(v)] + \frac{d}{dt^1} [\lambda'(v)\ddot{v}_{11} + \frac{\alpha}{3} \lambda^3(v) - \frac{1}{2} \lambda''(v)\dot{v}_1^2]$$
$$+ \dot{v}_1^3[\frac{\alpha}{3} \lambda'(v) + \frac{1}{2} \lambda'''(v)] .$$

But it follows directly by differentiation of (4.38) that the coefficient of $\dot{v}_1^3$ vanishes. Thus the condition that $\chi = 0$ implies that there exists a function U such that

(4.42) $$\dot{U}_1 = \lambda(v), \quad \dot{U}_2 = -\lambda'(v)\ddot{v}_{11} - \frac{\alpha}{3} \lambda^3(v) + \frac{1}{2} \lambda''(v)\dot{v}_1^2 .$$

From (4.18) we infer that $U = u + \theta(t_2)$; we choose $\theta(t_2)$ to be zero, so that $\dot{U}_2 = \dot{u}_2$. Then, for the case that $\alpha > 0$, we obtain from (4.42) and (4.39),

$$\dot{u}_2 = -a[\beta \ddot{v}_{11} \cos \beta v + \frac{\alpha}{3} (\dot{u}_1^2 + \dot{v}_1^2) \sin \beta v] , \tag{4.43}$$

where $\beta = \sqrt{2\alpha/3}$, while, when $\alpha < 0$,

$$\dot{u}_2 = -a[\beta \ddot{v}_{11} \cosh \beta v + \frac{\alpha}{3} (\dot{u}_1^2 + \dot{v}_1^2) \sinh \beta v] , \tag{4.44}$$

where $\beta = \sqrt{-2\alpha/3}$. _Thus (4.43) and (4.44) represent the second components of our Bäcklund transformations corresponding to the cases_ $\alpha > 0$ _and_ $\alpha < 0$, _respectively_. When $\alpha = 6$, $\beta = 2$, the relations (4.39a) and (4.43) reduce to the forms given by Lamb [3].

3. $n = 4$

When an attempt is made to discuss the case $n = 4$ along precisely the same lines as developed above, it would appear that it is not possible to find conditions which will ensure that $L(y) - L(x)$ is a divergence; this may imply that there are no variational Bäcklund transformations when $n = 4$. Therefore, guided by the analysis of Section III, one would be inclined to look instead for simple Bäcklund transformations for this case. Indeed, it is immediately evident from our construction that the Bäcklund transformations corresponding to the cases $n = 2$ and $n = 3$ ensure the vanishing of the quantity χ which appears in (4.25): hence these transformations are, in fact, also simple.

Thus, in dealing with the case $n = 4$, we shall begin with (4.25), for which we have

$$\chi = \dddot{v}_{111} + \dot{v}_2 + \alpha \dot{u}_1 \dot{v}_1 (\dot{u}_1^2 + \dot{v}_1^2) . \tag{4.45}$$

Again we seek a function $\lambda(v)$ such that, with $\dot{u}_1 = \lambda(v)$, a multiple of χ becomes a divergence. Since

$$\lambda'(v)\chi = \frac{d}{dt^2} [\lambda(v)] + \frac{d}{dt^1} [\lambda'(v)\ddot{v}_{11} + \frac{\alpha}{4} \lambda^4(v) - \frac{1}{2} \lambda''(v)\dot{v}_1^2]$$
$$+ \dot{v}_1^3 [\frac{1}{2} \lambda'''(v) + \alpha\lambda(v)\lambda'(v)] , \tag{4.46}$$

it follows that this requirement is met provided that $\lambda(v)$ satisfies the differential equation

(4.47) $$\lambda'''(v) + 2\alpha\lambda(v)\lambda'(v) = 0 \; .$$

The latter can be integrated directly to yield

(4.48) $$\lambda''(v) + \alpha\lambda^2(v) = C \; ,$$

where C is a constant. Thus, whenever $\lambda(v)$ is a solution of (4.48), it follows from (4.46) that the condition $\chi = 0$ is equivalent to the stipulation that there exists a function U such that

(4.49) $$\dot{U}_1 = \lambda(v), \quad \dot{U}_2 = -\lambda'(v)\ddot{v}_{11} - \frac{\alpha}{4}\lambda^4(v) + \frac{1}{2}\lambda''(v)\dot{v}_1^2 \; .$$

Again we infer that $\dot{U}_1 = \dot{u}_1$ or $U = u + \theta(t_2)$, where $\theta(t_2)$ is an arbitrary function which we choose to be zero. Hence (4.49) yields

(4.50) $$\dot{u}_2 = -\lambda'(v)\ddot{v}_{11} - \frac{\alpha}{4}\lambda^4(v) + \frac{1}{2}\lambda''(v)\dot{v}_1^2 \; .$$

We now have to distinguish between two distinct cases. When the constant C on the right side of (4.48) vanishes, the solution of the latter is

(4.51) $$\lambda(v) = -6\alpha^{-1}v^{-2} \; ,$$

and thus, by virtue of (4.50), <u>the resulting simple Bäcklund transformation is given by</u>

(4.52) $$\dot{u}_1 = -6\alpha^{-1}v^{-2}, \quad \dot{u}_2 = 12\alpha^{-1}v^{-3}\ddot{v}_{11} - \frac{\alpha}{4}\dot{u}_1^4 - 18\alpha^{-1}v^{-4}\dot{v}_1^2 \; .$$

However, when the constant C on the right side of (4.48) does not vanish, we integrate the latter, obtaining

(4.53) $$[\lambda'(v)]^2 + \frac{2\alpha}{3}\lambda^3(v) - 2C\lambda(v) - k = 0 \; ,$$

where k is a constant. From the theory of elliptic functions [9,p. 437] it follows that the solution of this equation is given by

(4.54) $$\lambda(v) = -\wp\left(\sqrt{\frac{\alpha}{6}}\,v + b\right) \; ,$$

where P denotes the Weierstrass elliptic function with invariants

(4.55) $$g_2 = 12\alpha^{-1}C, \quad g_3 = 6k\alpha^{-1} ,$$

and where b is a constant. It therefore follows that the corresponding <u>simple Bäcklund transformation is given by</u>

(4.56) $$\dot{u}_1 = -P(\sqrt{\tfrac{\alpha}{6}}\, v + b) ,$$
$$\dot{u}_2 = \sqrt{\tfrac{\alpha}{6}} P'(\sqrt{\tfrac{\alpha}{6}}\, v + b)\ddot{v}_{11} - \tfrac{\alpha}{4} P^4(\sqrt{\tfrac{\alpha}{6}}\, v + b) - \tfrac{\alpha}{12} P''(\sqrt{\tfrac{\alpha}{6}}\, v + b)\dot{v}_1^2 .$$

V. POSSIBLE GENERALIZATIONS

Instead of considering the Lagrangian (4.1), one might be inclined to investigate a more general Lagrangian of the type

(5.1) $$L = \tfrac{1}{2}\dot{x}_1\dot{x}_2 - \tfrac{1}{2}(\ddot{x}_{11})^2 + g(\dot{x}_1) ,$$

where $g(\dot{x}_1)$ is an as yet arbitrary differentiable function of $\dot{x}_1$. Proceeding as before, we obtain

(5.2) $$E(x) = \frac{d}{dt^1}[E(x)] ,$$

where

(5.3) $$E(x) \equiv \dot{x}_2 + g'(\dot{x}_1)\,\dddot{x}_{111} .$$

A direct criterion for the existence of simple Bäcklund transformations associated with the equation $E(x) = 0$ is given by the following

<u>Theorem</u>: <u>The partial differential equation</u>

(5.4) $$\dot{x}_2 + g'(\dot{x}_1) + \dddot{x}_{111} = 0$$

<u>possesses a simple Bäcklund transformation, if, for the given function</u> $g(\dot{x}_1)$, <u>there exists a pair of functions</u> $\phi(v)$, $\theta(v)$, <u>with</u> $\theta'(v) \neq 0$, <u>such that the following condition is satisfied:</u>

(5.5) $$g'(\dot{y}_1) - g'(\dot{x}_1) = -\frac{\theta'''(v)}{\theta'(v)}\dot{v}_1^3 - \frac{2\phi'(v)}{\theta'(v)}\dot{v}_1 .$$

<u>Under these conditions the corresponding simple Bäcklund transformations possess the form</u>

(5.6) $$\dot{u}_1 = \theta(v) ,$$

(5.7) $$\dot{u}_2 = -\theta'(v)\ddot{v}_{11} + \frac{1}{2}\theta''(v)\dot{v}_1^2 + \phi(v) .$$

<u>Proof:</u> From (4.25) and (5.3) we have

$$2\chi = 2(\dot{v}_2 + \dddot{v}_{111}) + g'(\dot{y}_1) - g'(\dot{x}_1) .$$

If it is now assumed that (5.5) is satisfied, this gives rise to

(5.8) $$\theta'(v)\chi = \theta'(v)(\dot{v}_2 + \dddot{v}_{111}) - \frac{1}{2}\theta'''(v)\dot{v}_1^3 - \phi'(v)\dot{v}_1$$

$$= \frac{d}{dt^1}[\theta'(v)\ddot{v}_{11} - \frac{1}{2}\theta''(v)\dot{v}_1^2 - \phi(v)] + \frac{d}{dt^2}[\theta(v)] .$$

By definition, the existence of a simple Bäcklund transformation relating the functions x and y must imply that $\chi = 0$; because of (5.8) this is tantamount to the existence of a function U which is such that

$$\dot{U}_1 = \theta(v), \quad \dot{U}_2 = -\theta'(v)\ddot{v}_{11} + \frac{1}{2}\theta''(v)\dot{v}_1^2 + \phi(v) .$$

The identification of u with U immediately gives rise to the relations (5.6) and (5.7).

<u>Remarks:</u> Needless to say, the Bäcklund transformations discussed in Section IV may be derived from the above theorem, if we put

(5.9) $$g(\dot{x}_1) = \frac{\alpha}{n(n+1)}(\dot{x}_1)^{n+1} ,$$

with $n = 2, 3, 4$. However, this immediately raises the question as to whether or not the cases $n \geq 5$ are also amenable to treatment by this method. A little reflection shows that this does not appear to be the case. For instance, when $n = 5$, we have from (5.9),

$$(5.10) \qquad g'(\dot{y}_1) - g'(\dot{x}_1) = \alpha[2\dot{u}_1^4\dot{v}_1 + 4\dot{u}_1^2\dot{v}_1^3 + \frac{2}{5}\dot{v}_1^5] ,$$

and when this is substituted in (5.5), it is evident that the requirement implied by (5.5) cannot be satisfied. Similar phenomena will obviously occur when $n > 5$.

Thus it would appear that the partial differential equation (4.11) does not possess simple Bäcklund transformations whenever $n \geq 5$. It should be noted, however, that the theorem merely yields sufficient conditions for the existence of Bäcklund transformations of the type (5.6), (5.7), and therefore does not necessarily preclude the existence of alternative types.

VI. THE HOPF-COLE TRANSFORMATION

We have thus far encountered three types of Bäcklund transformations, namely those which we have called variational, simple, and strong, respectively. The first of these is meaningful only when the partial differential equations under consideration are derivable from a variational principle; however, the following example will clearly indicate that the second and third are useful even when this condition is not met. To this end we shall consider the diffusion equation in conjunction with the Burgers equation, which we shall write respectively in the form

$$(6.1) \qquad D(x) = 0 ,$$

and

$$(6.2) \qquad B(x) = 0 ,$$

where

$$(6.3) \qquad D(x) \equiv \ddot{x}_{11} + \dot{x}_2 ,$$

and

(6.4) $$B(x) \equiv \ddot{x}_{11} + x\dot{x}_1 + \dot{x}_2 .$$

Clearly the relations

(6.5) $$y = x - \dot{U}_1,\ \dot{y}_1 = \dot{x}_1 - \frac{1}{2} y^2 + \dot{U}_2 ,$$

in which U denotes an arbitrary class C^2 function, represent a simple Bäcklund transformation, since they ensure that $B(y) = D(x)$.

The question now arises as to whether it is possible to choose the function U in (6.5) such that (6.5) is also a strong Bäcklund transformation in the sense that it ensures that both of the equations $D(x) = 0$ and $B(y) = 0$ are satisfied. This is easily achieved as follows. We observe that the validity of (6.1) implies the existence of a function V such that

(6.6) $$x = \dot{V}_1,\ \dot{x}_1 = -\dot{V}_2 .$$

Thus the first member of (6.5) can be expressed as

(6.7) $$y = \dot{W}_1 ,$$

where

(6.8) $$W \equiv V - U .$$

Substitution of (6.6), (6.7), and (6.8) in the second member of (6.5) then yields

(6.9) $$\ddot{W}_{11} + \frac{1}{2} \dot{W}_1^2 + \dot{W}_2 = 0 .$$

It is evident from our construction that any solution W of this equation will generate a strong Bäcklund transformation. Assuming for the moment that $W = W(x)$, we can write (6.9) in the form

$$(\dot{x}_1)^2[W_{xx} + \frac{1}{2} (W_x)^2] + (\ddot{x}_{11} + \dot{x}_2)W_x = 0 ,$$

which, because (6.1) is assumed to hold, gives rise to the solution $W_x(x) = 2x^{-1}$, or

(6.10) $$W = 2\ell n\ x .$$

When this is substituted in (6.7) we obtain

$$\dot{x}_1 = \frac{1}{2} xy \tag{6.11}$$

while (6.6) yields

$$\dot{x}_2 = -\frac{1}{2}(\dot{x}_1 y + x\dot{y}_1) . \tag{6.12}$$

Clearly (6.11) and (6.12) represent the well-known _Hopf-Cole transformation_ [8,p. 97] _which appears here as a strong Bäcklund transformation_.

However, the method sketched above leads to yet another such transformation. If we make the weaker assumption that $W = W(x,y)$, the condition (6.9) is satisfied whenever

$$W_{xx} + \frac{1}{2}(W_x)^2 = 0, \quad W_{yy} - \frac{1}{2}(W_y)^2 = 0, \quad W_{xy} = 0 ,$$

which gives rise to the solution

$$W = 2\,\ell n(xy^{-1}) . \tag{6.13}$$

The resulting alternative _strong Bäcklund transformation_ may then be written in the form

$$\dot{x}_1 = \frac{1}{2} xy + \frac{x}{y}\dot{y}_1, \quad \dot{x}_2 = -\frac{1}{4} xy^2 - \frac{1}{2} x\dot{y}_1 + \frac{x}{y}\dot{y}_2 . \tag{6.14}$$

Conversely, it may be verified directly that the integrability conditions of (6.14) imply that $B(y) = 0$, while differentiation of the first member of (6.14) with respect to t^1 leads to $D(x) = xy^{-1}B(y)$, thus confirming the theory.

VII. THE MIURA TRANSFORMATION

As a final application of the techniques discussed above, let us briefly consider the differential operators defined by

$$K_n(x) = \dot{x}_2 + \dddot{x}_{111} + \alpha x^{n-1}\dot{x}_1 , \tag{7.1}$$

where $n = 2,3,\ldots;$ clearly the case $n = 2$ corresponds to the Korteweg-deVries equation. Proceeding precisely as in the preceding section, we observe that the system

$$(7.2)\qquad y = x - \dot{U}_1, \quad \ddot{y}_{11} + \frac{1}{2}\alpha y^2 = \ddot{x}_{11} + \alpha n^{-1}x^n + \dot{U}_2 ,$$

is a simple Bäcklund transformation for the pair $K_2(y) = 0$, $K_n(x) = 0$, where U is again quite arbitrary. In order that (7.2) be strong, we require that U be such that (7.2) implies that $K_n(x) = 0$, in which case there exists a function V for which

$$(7.3)\qquad x = \dot{V}_1, \quad \ddot{x}_{11} + \alpha n^{-1}x^n = -\dot{V}_2 .$$

Thus, with $W = V - U$, the relations (7.2) yield

$$(7.4)\qquad y = \dot{W}_1, \quad \ddot{y}_{11} + \frac{1}{2}\alpha y^2 = -\dot{W}_2 ,$$

so that W must satisfy the condition

$$(7.5)\qquad \dddot{W}_{111} + \frac{1}{2}\alpha\dot{W}_1^2 + \dot{W}_2 = 0 ,$$

which, incidentally, possesses a formal structure identical with that of (4.11) when the latter is evaluated for $n = 2$. The equation (7.5) is the counterpart of (6.9) for the Hopf-Cole transformation, and we shall now seek solutions of the form $\dot{W}_1 = F(x,\dot{x}_1)$ for which (7.5) is satisfied subject to the condition $K_n(x) = 0$. A direct expansion of (7.5) in terms of F yields the conditions $F_{\dot{x}_1\dot{x}_1} = 0$, $F_{x\dot{x}_1} = 0$, which implies that we may write

$$(7.6)\qquad F(x,\dot{x}_1) = f(x) + k\dot{x}_1 + h ,$$

where k and h are constants. The substitution of this form of F in (7.5) then gives rise to the conditions

$$(7.7)\qquad 3f'' + \alpha k^2 = 0, \quad f = x^{n-1} - h ,$$

<u>which are consistent if and only if</u> $n \leq 3$. For $n = 3$, we thus obtain

$$(7.8)\qquad k^2 = -6\alpha^{-1} ,$$

and, putting $h = 0$, we infer from (7.7), (7.6), and (7.4) that

$$y = \dot{W}_1 = F = x^2 + k\dot{x}_1 . \tag{7.9}$$

This is the transformation which relates the modified Korteweg-deVries equation (since $n = 3$) to the Korteweg-deVries equation as given by Miura [5]. Moreover, from (7.8) and (7.9) it follows that

$$6(\ddot{x}_{11} + \frac{1}{3}\alpha x^3) = \alpha(2xy - k\dot{y}_1) ,$$

and this, together with (7.3), yields

$$\dot{x}_2 = -k\ddot{y}_{11} + 2k^{-2}(x\dot{y}_1 + \dot{x}_1 y) , \tag{7.10}$$

which must be adjoined to (7.9) in order to give rise to a strong Bäcklund transformation. (The pair (7.9), (7.10) is given by Lamb [3] for the case when $\alpha = 6$, or $k = \pm i$.)

Since this result is necessarily restricted to the case $n = 3$, one might be inclined to seek more general solutions of (7.5). However, a somewhat more laborious calculation based on a solution possessing the more general form $\dot{W}_1 = F(x,\dot{x}_1,\ddot{x}_{11})$ leads to the conclusion $\partial F/\partial\ddot{x}_{11} = 0$, which in turn entails all the previous restrictions. The above analysis therefore shows that <u>there are no strong Bäcklund transformations of the form</u> $y = F(x,\dot{x}_1,\ddot{x}_{11})$ <u>which relate the Korteweg-deVries equation to the equations</u> $K_n(x) = 0$ when $n > 3$. Analogous calculations yield negative conclusions of a similar nature for the pairs $K_n(x) = 0$, $K_p(x) = 0$ when $n > 3$, $p > n$.

REFERENCES

[1] J. ACZEL, Lectures on Functional Equations and Their Applications, Academic Press, London and New York, N. Y., 1966.

[2] G.L. LAMB, JR., Analytical descriptions of ultrashort optical pulse propagation in a resonant medium, Rev. Modern Phys. 43 (1971), 99-124.

[3] __________, Bäcklund transformation in nonlinear pulse propagation, Phys. Lett. 48A (1974), 73-74; Bäcklund transformations for certain nonlinear evolution equations, J. Mathematical Phys. 15 (1974), 2157-2165.

[4] D.W. MCLAUGHLIN AND A.C. SCOTT, A restricted Bäcklund transformation, J. Mathematical Phys. 14 (1973), 1817-1828.

[5] R.M. MIURA, Korteweg-deVries equation and generalizations. I. A remarkable explicit nonlinear transformation, J. Mathematical Phys. 9 (1968), 1202-1204.

[6] H. RUND, The Hamilton-Jacobi Theory in the Calculus of Variations, D. Van Nostrand, London and New York, 1966 (augmented and revised edition, Krieger Publishing Co., New York, 1973).

[7] H.D. WAHLQUIST AND F.B. ESTABROOK, Bäcklund transformation for solutions of the Korteweg-deVries equation, Phys. Rev. Lett. 31 (1973), 1386-1390.

[8] G.B. WHITHAM, Linear and Nonlinear Waves, Wiley-Interscience, New York, 1974.

[9] E.T. WHITTAKER AND G. WATSON, Modern Analysis, Cambridge University Press, Cambridge, 1940.

THE INTERRELATION BETWEEN BÄCKLUND TRANSFORMATIONS AND THE INVERSE SCATTERING TRANSFORM*

Alan C. Newell

Department of Mathematics
Clarkson College of Technology
Potsdam, New York 13676

I. INTRODUCTION AND GENERAL DISCUSSION

Substantial progress has been made in our understanding of nonlinear dispersive wave phenomena as a result of the discovery of the inverse scattering transform method for solving nonlinear evolution equations exactly. The method, first developed by Gardner, Greene, Kruskal, and Miura (GGKM) [1], associates with each nonlinear evolution equation an eigenvalue (scattering) problem in which the unknown variable in the evolution equation plays the role of a potential. Much work (Ablowitz, Kaup, Newell, and Segur (AKNS) [2]) has gone into identifying and classifying the broad range of evolution equations which can be associated with a given eigenvalue problem.

On the other hand, the "inverse" question as to how one chooses the appropriate eigenvalue problem for a given evolution equation is another matter and not yet resolved. One would like to have a systematic procedure by which one can build the appropriate eigenvalue problem from the equation of interest provided, of course, the equation is integrable (viewed as an infinite dimensional mechanical system). The most promising efforts in answering this question seem to be closely connected with a class of nonlinear transformations between solutions of the given equation or between solutions of the given equation and a closely related one. These transformations are called Bäcklund transformations after a specific transformation relating solutions of the sine-Gordon equation.

Indeed, it was just such a transformation which led GGKM to the Schrödinger equation. Miura [3] found that the transformation

*This work was partially supported by NSF Grants GA27727A1, GA32839X, and GP43653.

(1) $$u(x,t) = q^2(x,t) - iq_x(x,t)$$

related solutions of the Korteweg-deVries (KdV) equation,

(2) $$u_t + 6uu_x + u_{xxx} = 0 \ ,$$

and the modified Korteweg-deVries equation (MKdV),

(3) $$q_t + 6q^2q_x + q_{xxx} = 0 \ .$$

The linearization of the Riccati equation (1) immediately suggests the Schrödinger equation. Similarly, it was a transformation suggested by Kruskal [4] between solutions of

(4) $$u_{xt} = \sin u$$

and

(5) $$\frac{v_{xt}}{\sqrt{1-\epsilon^2 v_x^2}} = \sin v$$

which first led AKNS [5] to the correct choice of scattering problem for treating the sine-Gordon equation. At the same time, Lamb [6] used the already known Bäcklund transformation between solutions of (4) to arrive at the same eigenvalue problem in Schrödinger form.

So far, the only systematic methods for obtaining the Bäcklund transformation directly are those of Clairin [7] (extensively used and reported in detail in this volume by Lamb [8],[9]) and Wahlquist and Estabrook [10], also reported here [11]. However successful these methods have been in developing transformations for equations with well-known properties (like KdV), nothing substantially new has yet emerged. What one would really like is a straightforward method for deciding whether, for example, the equation

(6) $$u_{tt} - u_{xx} = F(u, \frac{\partial^n u}{\partial x^n})$$

for various functions F is integrable by the inverse scattering transform and, if it is, what scattering problem is the appropriate one. Nevertheless, the methods have many promising features and do attack one of the most important and intriguing questions in the general theory.

I might remark at this stage that there are other ad hoc ways of developing the appropriate eigenvalue problem [12]. They usually build from the dispersion relation of the linearized evolution equation.

The purpose of this short note is to develop the Bäcklund transformation by beginning from the eigenvalue problem. While the real interest in Bäcklund transformations is to have them provide the eigenvalue problem, it is nevertheless interesting to note the close interrelation. These ideas were first presented at the SIAM conference in October 1973 and are already present in some detail in the literature [2]. Chen [13], [14] has also developed these connections independently and applied them more extensively. In addition, we will point out that the Miura transformation (1) connects not only solutions of the Korteweg-deVries and modified Korteweg-deVries equations (2) and (3), but also connects solutions of each member of their respective families (each family being generated by taking in turn the conserved quantities as Hamiltonians) corresponding to the same dispersion relation. A few specific solutions are discussed and in particular the connection between the similarity solutions for (2) and (3) is shown.

II. DERIVATION OF BÄCKLUND TRANSFORMATION FROM THE EIGENVALUE PROBLEM AND VICE VERSA

The generalized Zakharov and Shabat eigenvalue problem is

$$\begin{aligned} \phi_{1x} + i\zeta\phi_1 &= q(x,t)\phi_2 , \\ \phi_{2x} - i\zeta\phi_2 &= r(x,t)\phi_1 , \end{aligned} \tag{7}$$

and the associated time-evolution equations are

$$\begin{aligned} \phi_{1t} &= (A-iA_0(\zeta))\phi_1 + B\phi_2 , \\ \phi_{2t} &= C\phi_1 - (A+iA_0(\zeta))\phi_2 . \end{aligned} \tag{8}$$

The class of evolution equations which can be handled by (7) and (8) may be written

$$\begin{pmatrix} r_t \\ -q_t \end{pmatrix} + 2iA_0(L^A)\begin{pmatrix} r \\ q \end{pmatrix} = 0 \tag{9}$$

where

$$L^A \equiv \frac{1}{2i}\begin{bmatrix} \frac{\partial}{\partial x} - 2r\int_{-\infty}^{x} dy\, q & 2r\int_{-\infty}^{x} dy\, r \\ -2q\int_{-\infty}^{x} dy\, q & -\frac{\partial}{\partial x} + 2q\int_{-\infty}^{x} dy\, r \end{bmatrix} . \tag{10}$$

Equation (9) results from the integrability condition obtained by cross-differentiation of (7) and (8). Its linearization immediately reveals the connection between $A_0(\zeta)$ and the dispersion relations for the r and q equations, respectively. Equations (7) and (8) are equivalent to the pair of Riccati equations for $\gamma \equiv \phi_2/\phi_1$:

$$\gamma_x = 2i\zeta\gamma + r - q\gamma^2 , \tag{11}$$

$$\gamma_t = -2A\gamma + C - B\gamma^2 . \tag{12}$$

<u>Remark</u>. If one tries to decouple (11) into a second-order linear system and demands that all coefficients be regular (inverse powers of q will be singular as $|x| \to \infty$), then one finds that (7) is the unique nonsingular choice.

The pair of Riccati equations (11), (12) lead directly to the Bäcklund transformation.

First we take $r \equiv -q \equiv \frac{1}{2}u_x$. Then define

$$\gamma \equiv \tan\frac{u+v}{4} \quad \text{or} \quad v \equiv -u + 4\tan^{-1}\gamma \tag{13}$$

and (11) becomes

$$v_x - u_x = 4i\zeta\,\frac{u+v}{2} . \tag{14}$$

We note that (14) can be inverted to (11) with u replaced by v and ζ by $-\zeta$.

Beginning with the vacuum solution $u = 0$, (14) generates a one-parameter family of solutions, each member corresponding to a soliton with complex characteristic parameter ζ. Lamb [15] has shown how multisoliton solutions can be obtained without further integration.

The structure of the x-component of the Bäcklund transformation is determined once the choice $r = -q = \frac{1}{2}u_x$ is made. The solutions of (3) and (4) (both are members of the same family) are interrelated by (14). The time part depends on the individual equation (characterized by its dispersion relation) in each family and on the corresponding choices of A, B, C in (8).

Since both $u(x,t)$ and $v(x,t)$ of (13) and (14) satisfy the same equation, each flow field shares in common the functionals which are the motion invariants. These quantities are all generated from the scattering function $a(\zeta) = \lim_{x\to\infty} \phi_1(\zeta,x)e^{i\zeta x}$ known as the transmission coefficient[1] and the first two are

$$I_1[u] = \frac{1}{8i}\int_{-\infty}^{\infty} u_x^2 dx ,$$

$$I_2[u] = \frac{1}{8i}\int_{-\infty}^{\infty} (-u_{xx}^2 + \tfrac{1}{4}u_x^4)\, dx .$$

Contrary to what has been suggested in the literature before, the conserved quantities map into each other in a 1-1 fashion, a feature distinctly different from the property of the Miura transformation which relates solutions of two different equations. In that case, one conserved quantity of the (slightly-modified) MKdV equation generates the infinite set of conserved quantities for the KdV equation. For example, let $v(x,t)$ be a $(2n+2)\pi$-pulse $(v(+\infty)-v(-\infty) = (2n+2)\pi)$ generated from the $2n\pi$-pulse $u(x,t)$ by (14) with $\zeta = \bar{\zeta} = i\bar{\eta}$. Then, from (14),

$$\int_{-\infty}^{\infty} (v_x - 2i\bar{\zeta} \sin \frac{u+v}{2})^2 dx = \int_{-\infty}^{\infty} (u_x + 2i\bar{\zeta} \sin \frac{u+v}{2})^2 dx ,$$

from which

[1] In the usual connotation, this function is the inverse transmission coefficient.

(15) $$I_1[v] = I_1[u] + 2i\bar{\eta} .$$

Since for $0 < |\mathrm{Arg}\zeta| < \pi$, the asymptotic expression for $\ell n\ a[v]$ is $\frac{1}{\zeta^n} \sum_{n=1}^{\infty} I_n[v]$, (15) is consistent with

(16) $$a[v] = a[u] \frac{\zeta+i\bar{\eta}}{\zeta-i\bar{\eta}} .$$

One might therefore conjecture that if $u(x,t)$ is any solution of the evolution equation with sufficiently smooth properties and sufficient decay as $|x| \to \infty$ to ensure the definition and existence of the scattering functions, the effects of (14) on the scattering data are to add one bound state and to change the signs in the arguments. (Recall that $v(x,t)$ is the potential in (7) with $\zeta \to -\zeta$.)

A specific time dependency is added to (14) by specifying the particular evolution equation belonging to the family for which (7) with $r = -q$ is the appropriate eigenvalue problem. For the sine-Gordon equation, $A = \frac{i}{4\zeta}\cos u$, $B = C = \frac{i}{4\zeta}\sin u$ and from (12) and (13), we have

(17) $$\frac{v_t+u_t}{4} = \frac{i}{4\zeta} \sin \frac{u-v}{2} = \frac{i}{4(-\zeta)} \sin \frac{v-u}{2} .$$

For this case, the fact that $v(x,t)$ also satisfies (4) can be verified by cross-differentiation of (14) and (17). This verification is not so straightforward for the MKdV equation (3) for which $A = -4i\zeta^3 + 2i\zeta q^2$, $B = 4q\zeta^2 + 2i\zeta q_x - q_{xx} - 2q^3$, $C = -4q\zeta^2 + 2i\zeta q_x + q_{xx} + 2q^3$ and from (12) and (13)

(18) $$\frac{v_t+u_t}{4} = (4i\zeta^3-2iq^2\zeta)\sin \frac{u+v}{2} + 2iq_x\zeta \cos \frac{u+v}{2} + q_{xx} - 4q\zeta^2 + 2q^3$$

where again $q = -\tfrac{1}{2}u_x$. Now, if $p \equiv -\tfrac{1}{2}v_x$, then it can be shown, using (14), that the right side of (18) is equal to the same expression with q, u, and ζ interchanged with p, v, and $-\zeta$, respectively.

Remark. The proof that $v(x,t)$ (and $p(x,t)$) satisfies the same evolution equation as $u(x,t)$ (and $q(x,t)$) is obvious once we establish the symmetry property of (14) and (18). Then all steps may be retraced to (7) and (8) except that q, u, and ζ are replaced by p, v, and $-\zeta$. But the integrability condition of (7) and (8) is just the evolution equation.

The Bäcklund transformations corresponding to (14) and (17), and (14) and (18) may be obtained by similar calculations for the other members of the $r = -q$ family.

For $r \equiv q \equiv \frac{1}{2}u_x$, the appropriate analogue to (13) is

$$\gamma = \tanh \frac{u+v}{4} , \tag{19}$$

whence

$$v_x - u_x = 4i\zeta \sinh \frac{u+v}{2} . \tag{20}$$

A particular evolution equation which belongs to this family is the sinh-Gordon equation

$$u_{xt} = \sinh u \tag{21}$$

for which $A = \frac{i}{4\zeta} \cosh u$, $-B = C = \frac{i}{4\zeta} \sinh u$ and using (12) and (19),

$$v_t + u_t = \frac{i}{\zeta} \sinh \frac{u-v}{2} . \tag{22}$$

Again, similar calculations may be performed for other members of this family.

For $r \equiv -1$, $q = q(x,t)$, it is convenient to define $\gamma \equiv \phi_1/\phi_2$ and we find,

$$\gamma_x = -2i\zeta\gamma + q + \gamma^2 ,$$

which may be written

$$(\gamma-i\zeta)_x = (\gamma-i\zeta)_x^2 + q + \zeta^2 . \tag{23}$$

Note that this is the Miura transformation (1). If q satisfies KdV, then

$$\gamma_t = (-8i\zeta^3+4iq\zeta-2q_x)\gamma + (4\zeta^2 q+2iq_x\zeta-q_{xx}-2q^2) - (2q-4\zeta^2)\gamma^2 . \tag{24}$$

Defining $\gamma-i\zeta \equiv \frac{1}{2}(u'-u)$, $q \equiv -u_x$, $q' \equiv -u'_x$, we obtain the Bäcklund transformation,

(25) $$\frac{u'_x+u_x}{2} = \left(\frac{u'-u}{2}\right)^2 + \zeta^2 \ ,$$

(26) $$\frac{u'_t+u_t}{2} = -2q_x\left(\frac{u'-u}{2}\right) - 2q\left(\frac{u'-u}{2}\right)^2 + q^2 - 2\zeta^2 q' \ .$$

The fact that q' satisfies KdV is proven by noting that, with the help of (25), (26) may be written as the same expression with u, q, and ζ interchanged with u', q', and ζ, respectively.

By similar procedures, one may develop Bäcklund transformations for any members of the class of evolution equations (9) for which $r = -q$, $r = q$, or $r = -1$. The eigenvalue problem when $r = -1$ becomes the Schrödinger equation

(27) $$\phi_{2xx} + (\zeta^2 + q(x,t))\phi_2 = 0 \ .$$

The evolution equations associated with (27) can be characterized

(28) $$q_t + C_0(L_S^A)q_x = 0$$

where

(29) $$L_S^A[q] \equiv -\frac{1}{4}\frac{\partial^2}{\partial x^2} - q + \frac{1}{2}q_x\int_x^\infty dy$$

and $C_0(\zeta^2)$ is directly connected to the phase speed of the linearized waves of (28) (see [2]).

We will now show that the Miura transformation is immediately recovered by considering (7) and (27) and that further it relates all the solutions of each member of the $r = \mp q$ families with the corresponding (having the same dispersion relation) equation in the $r = -1$ family.

III. THE MIURA TRANSFORMATION

Consider (7) with $r = -q$. We can readily show that

(30) $$(\phi_1 - i\phi_2)_{xx} + (\zeta^2 + q^2 - iq_x)(\phi_1 - i\phi_2) = 0 \ ,$$

from which we know that if $q^2 - iq_x$ satisfies the KdV equation the spectrum of

(30) is preserved, an observation which very naturally suggests the Miura transformation

(31) $$u(x,t) = q^2(x,t) - iq_x(x,t) \ .$$

We now show that all members of the $r = -q$ family are related to the corresponding member of the KdV family by the <u>same</u> transformation.

First we note that if $r = -q$, the general evolution euuation (9) may be written (see [2])

(32) $$q_t + D_0(L^A)q_x = 0$$

where

(33) $$L^A[q] \equiv -\frac{1}{4}\frac{\partial^2}{\partial x^2} - q^2 - q_x \int_{-\infty}^{x} dy\ q$$

and $D_0(\zeta^2)$ is related to the phase speed of linear waves. Define the operator,

(34) $$\hat{L}^A[q] \equiv L^A[q] + q_x \int_{-\infty}^{\infty} dy\ q \ .$$

Then for arbitrary $v(x,t)$,

(35) $$\hat{L}^A[q]\cdot v = L^A[q]\cdot v \quad \text{iff} \quad \int_{-\infty}^{\infty} qv\ dx = 0 \ .$$

We now state the result that

(36) $$L_S^A[q^2-iq_x]\cdot(2q-i\frac{\partial}{\partial x})\cdot v(x,t) \equiv (2q-i\frac{\partial}{\partial x})\cdot\hat{L}^A[q]\cdot v(x,t)$$

for all sufficiently smooth $v(x,t)$. The proof may be carried through by direct calculation using (29) and (34). It follows immediately that

(37) $$\{L_S^A[q^2-iq_x]\}^n\cdot(2q-i\frac{\partial}{\partial x})\cdot v \equiv (2q-i\frac{\partial}{\partial x})\{\hat{L}^A[q]\}^n\cdot v$$

for any integer n. Thus, it follows that if $C_0(\zeta^2)$ may be written as a series in ζ^2 and $u(x,t) = q^2 - iq_x$, then

(38) $$u_t + C_0(L_S^A)u_x = (2q-i\frac{\partial}{\partial x})(q_t + C_0(\hat{L}^A)q_x) \ .$$

But $(\hat{L}^A)^n q_x = (L^A)^n q_x$ which can be proven by direct calculation. More simply, one may notice that since $\int_{-\infty}^{\infty} q^2 dx$ is a motion invariant for each member of the equation class (32), $\int_{-\infty}^{\infty} q(L^A)^n q_x dx = 0$ for all $n = 0,1,\ldots$. From (35), this is the necessary and sufficient condition that $\hat{L}^A(\hat{L}^A)^n q_x = L^A(L^A)^n q_x$ for each $n = 0,1,\ldots$. Hence, if $C_0(\zeta^2)$ is an entire function of ζ^2, then $C_0(\hat{L}^A)q_x = C_0(L^A)q_x$ and

(39) $$u_t + C_0(L_S^A)u_x = (2q - i\frac{\partial}{\partial x})(q_t + C_0(L^A)q_x) ,$$

which is the generalization of Miura's formula for the interrelation of equations (1) and (2).

Furthermore, if $C_0(\zeta^2)$ is singular, equal to $\frac{T(\zeta^2)}{S(\zeta^2)}$ say, then it may be shown that

(40) $$S(L_S^A)u_t + T(L_S^A)u_x = (2q - i\frac{\partial}{\partial x})\{S(L^A)q_t + T(L^A)q_x\}$$

where we must demand that: (a) any zeros of S do not belong to the discrete spectrum of the Schrödinger or Zakharov and Shabat operators and (b) if any zeros of S lie in the continuous spectrum, the initial data must be such that the scattering function known as the reflection coefficient is zero at the singular point. For other discussions on this point see [2] and, more recently, [16].

The $r = q$ case is analogous. The corresponding class of evolution equations may be written

(41) $$q_t + D_0(M^A)q_x = 0$$

where

(42) $$M^A[q] = -\frac{1}{4}\frac{\partial^2}{\partial x^2} + q^2 + q_x\int_{-\infty}^{x} dy\, q .$$

Again, rewriting (7), we may show

(43) $$(v_1 - v_2)_{xx} + (\zeta^2 + q_x - q^2)(v_1 - v_2) = 0$$

which suggests the transformation

$$u(x,t) = q_x - q^2 ,$$

which is exactly (31) with q replaced by iq. Obviously, therefore

$$L_S^A[q_x - q^2]\cdot\left(\frac{\partial}{\partial x} - 2q\right)\cdot v = \left(\frac{\partial}{\partial x} - 2q\right)\cdot \hat{M}^A[q]v \tag{44}$$

where

$$\hat{M}^A v = M^A v - q_x \int_{-\infty}^{\infty} qv\, dx .$$

For the class of v's considered, $\int_{-\infty}^{\infty} qv\, dx = 0$ and $\hat{M}^A v = M^A v$. Thus

$$u_t + C_0(L_S^A)u_x = \left(\frac{\partial}{\partial x} - 2q\right)(q_t + C_0(M^A)q_x) . \tag{45}$$

IV. PARTICULAR SOLUTIONS

There is a close connection, then, between members of the family (28) and members of the family (32) with the same dispersion relation $(D_0 = C_0)$. If q satisfies (32), then from (39) u satisfies (28). The inverse is more complicated. If u satisfies (25), then from (31) and (39), there is a one-parameter family of corresponding q's which satisfy

$$q_t + 6q^2 q_x + q_{xxx} = 0$$

and for which $q \to 0$ as $|x| \to \infty$. For example, if $u(x,t) = 2K^2 \mathrm{sech}^2 K(x-4K^2t)$, the soliton solution of (1), then the general solution of the Riccati equation[2] (31) is

$$q(x,t) = \frac{-2iK}{\sinh 2K(x-4K^2t)} + \frac{\coth^2 K(x-4K^2t)}{A_0 + i(x - \frac{1}{K}\coth K(x-4K^2t))} , \tag{46}$$

where A_0 is an arbitrary constant. We note a number of interesting features of this solution. Firstly, if $A_0 = \infty$, the solution is singular at $x = 4K^2t$ but does decay at ∞. Secondly, if $A_0 \neq \infty$, there is no singularity (the singularities at $x = 4K^2t$ in both terms cancel) and $q \sim \frac{1}{x}$ for large x. If $q \to iq$,

[2] I am grateful to E. Barouch for a useful suggestion here.

(46) provides a local (but not absolutely integrable) real solution for the MKdV equation

$$q_t - 6q^2 q_x + q_{xxx} = 0 ,$$

for which there are no solitons (no bound states of the Zakharov and Shabat operator) and no simply translating solitary waves. The role of such solutions in the general picture or even their stability has not yet been investigated.

Finally, we remark on the interrelation between the similarity solutions for the equation classes (28) and (32). It is convenient to write (32) in the form (see Flaschka and Newell [16])

$$q_t = i \frac{\partial}{\partial x} (F(L_F^A) q) \tag{47}$$

where $F = F(\zeta^2)$ and

$$L_F^A[q,x] = -\frac{1}{4}\frac{\partial^2}{\partial x^2} - q^2 + q \int_{-\infty}^{x} dy \; q_x . \tag{48}$$

Now it can be shown

$$L_F^A[q,x] v(\frac{x}{t^\alpha}) = \frac{1}{t^{2\alpha}} L_F^A[g,\eta] v(\eta) \tag{49}$$

where

$$q = \frac{1}{t^\alpha} g(\eta) , \quad \eta = \frac{x}{t^\alpha} . \tag{50}$$

If $F(L_F^A) = ic(L_F^A)^n$, the similarity solutions $g(\eta)$ satisfy an equation which is once integrable to

$$\alpha\eta g + c(L_F^A[g,\eta]^n) g = 0 , \quad \alpha = \frac{1}{3n} . \tag{51}$$

On the other hand, the similarity solutions for the equation class (28), with $C_0(\zeta^2) = (\zeta^2)^n$ are

$$u(x,t) = \frac{1}{t^{2\alpha}} f(\eta) = \frac{1}{t^{2\alpha}} (g^2 - ig') , \quad \alpha = \frac{1}{3n} . \tag{52}$$

In the case where $n = 1$, the similarity solutions of the KdV equations are found from (52) and (51) which is now $(c = -4)$

$$\frac{1}{3}\eta g + g'' + 2g^3 = 0 , \tag{53}$$

the well-known Painleve equation.[3]

REFERENCES

[1] C.S. GARDNER, J.M. GREENE, M.D. KRUSKAL, AND R.M. MIURA, Method for solving the Korteweg-deVries equation, Phys. Rev. Lett. 19 (1967), 1095-1097; Korteweg-deVries equation and generalizations. VI. Methods for exact solution, Comm. Pure Appl. Math. 27 (1974), 97-133.

[2] M.J. ABLOWITZ, D.J. KAUP, A.C. NEWELL, AND H. SEGUR, The inverse scattering transform - Fourier analysis for nonlinear problems, Studies in Appl. Math. 53 (1974), 249-315.

[3] R.M. MIURA, Korteweg-deVries and generalizations. I. A remarkable explicit nonlinear transformation, J. Mathematical Phys. 9 (1968), 1202-1204.

[4] M.D. KRUSKAL, The Korteweg-deVries equation and related evolution equations, Nonlinear Wave Motion, A.C. Newell, Ed., Lectures in Applied Mathematics, Vol. 15, American Mathematical Society, Providence, Rhode Island, 1974, 61-83.

[5] M.J. ABLOWITZ, D.J. KAUP, A.C. NEWELL, AND H. SEGUR, Method for solving the sine-Gordon equation, Phys. Rev. Lett. 30 (1973), 1262-1264.

[6] G.L. LAMB, JR., Phase variation in coherent-optical-pulse propagation, Phys. Rev. Lett. 31 (1973), 196-199.

[7] J. CLAIRIN, Sur quelques équations aux dérivées partielles du second ordre, Ann. Fac. Sci. Univ. Toulouse, 5 (1903), 437-458.

[8] G.L. LAMB, JR., Bäcklund transformations for certain nonlinear evolution equations, J. Mathematical Phys. 15 (1974), 2157-2165.

[9] ______________, Bäcklund transformations at the turn of the century, this volume.

[10] H.D. WAHLQUIST AND F.B. ESTABROOK, Prolongation structures of nonlinear evolution equations, J. Mathematical Phys. 16 (1975), 1-7.

[11] H.D. WAHLQUIST, Bäcklund transformation of potentials of the Korteweg-deVries equation and the interaction of solitons with cnoidal waves, this volume.

[12] D.J. KAUP, Finding eigenvalue problems for solving nonlinear evolution equations, to be published.

[13] H.-H. CHEN, General derivation of Bäcklund transformations from inverse scattering problems, Phys. Rev. Lett. 33 (1974), 925-128.

---

[3] The simplifying transformation $f = g^2 - ig'$ was to my knowledge, first used by Whitham.

[14] ________, Relation between Bäcklund transformations and inverse scattering problems, this volume.

[15] G.L. LAMB, JR., Analytical descriptions of ultrashort optical pulse propagation in a resonant medium, Rev. Modern Phys. 43 (1971), 99-124.

[16] H. FLASCHKA AND A.C. NEWELL, Integrable systems of nonlinear evolution equations, Dynamical Systems, Theory and Applications, Battelle Seattle 1974 Rencontres, J. Moser, ed., Springer-Verlag, New York, N.Y., 1975, pp. 355-440.

RELATION BETWEEN BÄCKLUND TRANSFORMATIONS AND INVERSE SCATTERING PROBLEMS*

Hsing-Hen Chen†

Institute for Advanced Study
Princeton, New Jersey 08540

I. INTRODUCTION

It is now well-known [1] that a large class of nonlinear evolution equations exists with multisoliton solution structure. Lax [2] showed first the criterion for constructing such equations. Given two linear operators $L(x,t)$ and $A(x,t)$ satisfying the operator equation

$$[L,A] = -\partial_t L \ , \tag{1}$$

where $[\ ,\ ]$ is the commutator, the eigenvalue λ of L,

$$L\psi = \lambda\psi \ , \tag{2}$$

is independent of t if and only if the corresponding eigenfunction ψ evolves in t according to

$$A\psi = \partial_t \psi \ . \tag{3}$$

A standard example is the Korteweg-deVries (KdV) equation

$$q_t + 12qq_x + q_{xxx} = 0 \tag{4}$$

with

$$L \equiv \partial_x^2 + 2q(x,t) \tag{5}$$

$$A \equiv -4\partial_x^3 - 6(q\partial_x + \partial_x q) \ . \tag{6}$$

Gel'fand and Levitan [3] showed a long time ago how to construct the

*Work supported in part by AEC Grant AT(11-1)-3237.

†Present Address: Department of Physics and Astronomy, University of Maryland, College Park, Maryland 20742.

potential function q in the operator L from knowledge of the spectral function $\rho(\lambda)$. Kruskal et al [4] later demonstrated that for boundary conditions $q \to 0$ as $|x| \to \infty$, the equation can be solved by using the Gel'fand-Levitan-Marchenko equations. These methods combined into the inverse-scattering problem for non-linear evolution equations that is now famous.

On the other hand, there exists another method, the Bäcklund transformations (BT) [5], [6], [7], to obtain multisoliton solutions for nonlinear evolution equations. We will demonstrate in the following a general way of obtaining the Bäcklund transformation from the Lax equations. The reader may notice the simplicity of the derivation of BT compared to the involved inverse-scattering problem, especially for higher-order scattering problems.

II. BÄCKLUND TRANSFORMATION FOR THE KORTEWEG-DEVRIES EQUATION

From the Lax equations for the KdV equation,

$$\partial_x^2\psi + 2q\psi = \lambda\psi \ ,$$

(7)

$$4\partial_x^3\psi + 6(q\partial_x + \partial_x q)\psi = -\partial_t\psi \ ,$$

we are going to derive its BT. It is obvious that this set of equations represents a transformation between two functions $q(x,t)$ and $\psi(x,t)$. Substituting $u \equiv \psi_x/\psi$ into it, we get

(8a) $$u_x + u^2 + 2q = \lambda \ ,$$

(8b) $$4u_{xxx} + 12uu_{xx} + 12u^2u_x + 12u_x^2 + 12q_xu + 12qu_x + 6q_{xx} = -u_t \ .$$

Eliminating q, we observe that u satisfies

(9) $$u_t - 6u^2u_x + 6\lambda u_x + u_{xxx} = 0 \ ,$$

the modified Korteweg-deVries (MKdV) equation. Equation (8a) is therefore the famous Miura transformation between KdV and MKdV. Now, it is trivial to see that if u is a solution of (9), then $-u$ is also a solution. Correspondingly, we can find a q', a solution to the KdV equation, such that

$$-u_x + u^2 + 2q' = \lambda \tag{10a}$$

$$-4u_{xxx} + 12uu_{xx} - 12u^2u_x + 12u_x^2 - 12q'_xu - 12q'u_x + 6q'_{xx} = u_t \, . \tag{10b}$$

Subtracting (10a) from (8a), we get $u_x = q'-q$, or $u \equiv w'-w$ with $w_x \equiv q$. Substituting it back into (8) we then get the self-Bäcklund transformation for the KdV equation

$$(w'+w)_x = \lambda - (w'-w)^2 \tag{11a}$$

$$(w-w')_t = 4(w'-w)_{xxx} + 12(w'-w)(w'-w)_{xx} + 12(w'-w)^2(w'-w)_x + 12(w'-w)_x^2 + 12(w'-w)w_{xx} + 12w_x(w'-w)_x + 6w_{xxx} \, . \tag{11b}$$

The logical steps in obtaining the BT can be illustrated by the diagram

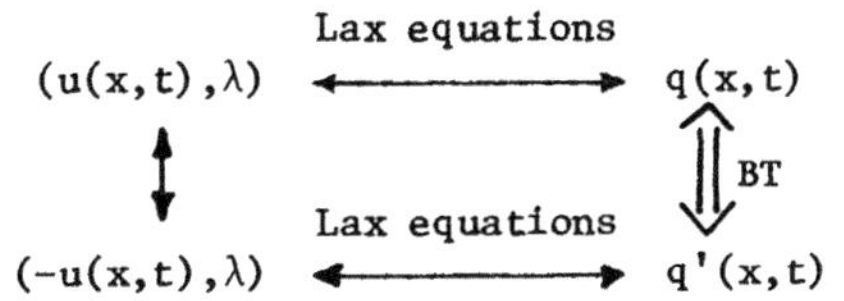

It is therefore clear that BT can be derived simply from the Lax equations. Soliton solutions will be constructed from this transformation. It is more convenient, however, to construct a superposition formula from (11) and thereby avoid the integration quadrature associated with the BT. The superposition formula was found by Wahlquist and Estabrook [6] to be

$$w_3 = w_0 + \frac{\lambda_1-\lambda_2}{w_1-w_2} \tag{12}$$

where w_0 is a known solution, w_1 and w_2 are solutions generated from w_0 by the parameters λ_1 and λ_2, respectively, and w_3 is a solution generated from w_1 by the parameter λ_2 or from w_2 by λ_1.

III. EXAMPLES OF BÄCKLUND TRANSFORMATIONS

In this section, we show the derivation of BT for nonlinear evolution equations constructed within the scheme of Ablowitz et al [8]. Ablowitz et al considered a particular 2×2 system of coupled Lax equations to describe the MKdV, sine-Gordon, and nonlinear Schrödinger equations, i.e.

$$v_{1x} + i\zeta v_1 = qv_2 \ , \qquad (13)$$
$$v_{2x} - i\zeta v_2 = rv_1 \ ,$$

and

$$v_{1t} = Av_1 + Bv_2 \ , \qquad (14)$$
$$v_{2t} = Cv_1 - Av_2 \ ,$$

with the integrability conditions

$$A_x = qC - rB \ ,$$
$$B_x = q_t - 2Aq - 2i\zeta B \ , \qquad (15)$$
$$C_x = r_t + 2Ar + 2i\zeta C \ .$$

Finite series expansions of A, B, and C in terms of ζ reduce the problems to specific equations of interest. Letting $u \equiv v_1/v_2$, we get two Riccati equations from (13) and (14).

$$u_x = -2i\zeta u - ru^2 + q \ , \qquad (16)$$
$$u_t = 2Au - Cu^2 + B \ .$$

This coupled set can be considered a transformation between $(q,r) \leftrightarrow (u,\zeta)$. We can furthermore divide the problem into different classes.

1. Class I: $r = \text{constant} = -2,\ i\zeta = k$

Equation (16) then becomes

$$u_x = -2ku + 2u^2 + q\ ,$$
(17)
$$u_t = 2Au - Cu^2 + B\ .$$

The simplest example in this class is the KdV equation [6], $q_t + 12qq_x + q_{xxx} = 0$. Following Ablowitz et al, we identify A, B, and C to be

$$A = 4k^3 + 4kq - 2q_x\ ,$$
(18)
$$B = -4k^2q + 2kq_x - q_{xx} - 4q^2\ ,$$
$$C = 8k^2 + 8q\ .$$

If we eliminate q from (17) and (18) we get

(19) $$u_t - 24u^2u_x + 24kuu_x + u_{xxx} = 0\ .$$

Equation (17) provides a BT between solutions of the KdV equation and (19). In particular, if $k = 0$, (19) reduces to the pure MKdV equation and (17) becomes the Miura transformation. Now, we can see that if (u,k) satisfies (19), then so does $(-u,-k)$. This gauge-like invariance of (19) tells us immediately that a second solution q' exists for the KdV equation such that

$$-u_x = -2ku + 2u^2 + q'\ ,$$
(20)
$$-u_t = -2A(q',-k)u - C(q',-k)u^2 + B(q',-k)\ .$$

From (17) and (20), we get

(21a) $$2u_x = q - q'\ ,$$

(21b) $$q + q' = 4ku - 4u^2\ ,$$

and

$$2u_t = 2u(A+A') - u^2(C-C') + (B-B') ,$$

(22)

$$0 = 2u(A-A') - u^2(C+C') + (B+B') .$$

Two different forms of the BT can be derived from (20):

(i) Let $q \equiv w_x$, $q' \equiv w'_x$, we have then $u = \frac{1}{2}(w - w' + k)$ and we get the BT between w and w'

$$(w + w')_x = k^2 - (w' - w)^2 ,$$

(23)

$$(w - w')_t = 2A(w - w' + k) - \frac{C}{2}(w - w' + k)^2 + 2B .$$

This is identical to (11) if we identify $k^2 = \lambda$.

(ii) Solve for u in (21b), getting $u = \frac{1}{2}(k \pm \sqrt{k^2-(q+q')})$. Because (21b) is a local equation, we can choose a constant x_0 such that we can generate continuous soliton solutions from the vacuum, then

(24)
$$u = \tfrac{1}{2}[k \pm \sqrt{k^2-(q+q')}\ \mathrm{sgn}(x-x_0-4k^2t)]$$

where sgn( ) is the signum function. From (24), we have

$$q + q' = k^2 - (2u - k)^2 .$$

Taking the x-derivative, then substituting u_x from (21a), we get

(25a)
$$(q+q')_x = \mp 2 \sqrt{k^2-(q+q')}\ \mathrm{sgn}(x-x_0-4k^2t)(q-q') .$$

In conjunction with the first equation in (22),

(25b)
$$2u_t = 2u(A+A') - u^2(C-C') + (B-B') ,$$

we get a BT relating q to q' directly. It can also be shown that the second equation in (22) is not independent. It is equivalent to (21). The two forms of BT, (25) and (23), are also equivalent.

2. Class II: $r = -q,\ i\zeta = k/2$

In this class, we can also get two equivalent forms of the Bäcklund transformation. Let us consider two examples.

a) For the MKdV equation, $q_t + 6q^2 q_x + q_{xxx} = 0$, we have

$$A = \frac{k^3}{2} + kq^2 ,$$

$$B = -k^2 q + kq_x - q_{xx} - 2q^3 , \tag{26}$$

$$C = k^2 q + kq_x + q_{xx} + 2q^3 .$$

Eliminating q from (16) and (26), we get the following equation

$$u = \tan \frac{v}{2} ,$$

$$v_t + v_{xxx} + \frac{v_x^3}{2} + \frac{3k^2}{2} v_x \sin^2 v = 0 . \tag{27}$$

This equation possesses the invariances $(u,k) \to (\pm u, \pm k)$ and leads to the following self-Bäcklund transformations for the MKdV equation:

(i) $v = w \mp w'$

$$(w \pm w')_x = k \sin(w \mp w'),$$

$$(w \pm w')_t = -2kw_x^2 \sin(w \mp w') \mp 2kw_{xx} \cos(w \mp w') \tag{28}$$

$$\mp k^3 \sin(w \mp w') \pm 2k^2 w_x ,$$

(ii) $$u = \frac{k \pm \sqrt{k^2 - (q \mp q')^2}\ \mathrm{sgn}(x - x_0 - k^2 t)}{q \mp q'}$$

$$(q \mp q')_x = \mp (q \pm q') \sqrt{k^2 - (q \mp q')^2}\ \mathrm{sgn}(x - x_0 - k^2 t) ,$$

$$u_t = 2u(A + A') - u^2 (C - C') + (B - B') . \tag{29}$$

b) For the sine-Gordon equation, $2w_{xt} = \sin 2w$, we have

$$A = -(\cos 2w)/2k \ ,$$
(30)
$$B = C = q_t/k \ .$$

The equation satisfied by u is

$$u = \tan \frac{v}{2} \ ,$$
(31)
$$v_{xt} = \sqrt{1-k^2 v_t^2}\, \sin v \ .$$

It is also invariant under the transformation $(u,k) \to (\pm u, \pm k)$. The Bäcklund transformations are then

(i) $v = w \mp w'$

$$(w \pm w')_x = k \sin(w \mp w') \ ,$$
(32)
$$(w \mp w')_t = \frac{1}{k} \sin(w \pm w') \ .$$

(ii) same as a) (ii) above but with appropriate A, B, C, and sgn.

3. Class III: $r = -q^*$, $\zeta = \xi + i\eta$

In this class, we have

$$u_x = -2i\zeta u + q^* u^2 + q \ ,$$
(33)
$$u_t = 2Au - Cu^2 + B \ .$$

The simplest example is the nonlinear Schrödinger equation $iq_t + q_{xx} + 2q^2 q^* = 0$. For this equation, we have

$$A = -2i\zeta^2 + iqq^* \ ,$$
(34)
$$B = 2q\zeta + iq_x \ ,$$
$$C = -2q^*\zeta + iq^*_x \ .$$

After elimination of q and q^* in (33), we get a nonlinear partial differential equation for u and u^*. It is straightforward to show that this equation is invariant under the gauge-like transformation $(u,\zeta) \to (\pm u, \zeta^*)$. Therefore, we have

$$(35)\qquad \begin{aligned} \pm u_x &= \mp 2i\zeta^* u + q'^* u^2 + q' , \\ \pm u_t &= \pm 2A(q',\zeta^*)u - C(q',\zeta^*)u^2 + B(q',\zeta^*) . \end{aligned}$$

From (33) and (35) we have

$$(36a)\qquad 0 = 4\eta u + (q \mp q')^* u^2 + (q \mp q') ,$$

$$(36b)\qquad 2u_x = -4i\xi u + (q \pm q')^* u^2 + (q \pm q') .$$

Equation (36b) is not integrable by the transformation $q = w_x$ as in the case of the KdV and MKdV equations. We therefore get only one form of self-Bäcklund transformation for the nonlinear Schrödinger equation:

$$(37)\qquad u = \frac{-2\eta \pm \sqrt{4\eta^2 - |q \mp q'|^2}\ \mathrm{sgn}(x - x_0 + 4\xi t)}{(q \mp q')^*} ,$$

$$(38)\qquad \begin{aligned} (q \pm q')_x &= -2i\xi(q \pm q') + (q \mp q')\sqrt{4\eta^2 - |q \pm q'|^2}\ \mathrm{sgn}(x - x_0 + 4\xi t) , \\ (q \pm q')_t &= -2i\xi(q \pm q')_x + i(q \mp q')_x \sqrt{4\eta^2 - |q \pm q'|^2}\ \mathrm{sgn}(x - x_0 + 4\xi t) \\ &\quad + i(q \pm q')(qq^* + q'q'^*) . \end{aligned}$$

We remark that Class II is only a special subclass of Class III, i.e. when $\zeta = -ik/2$ is pure imaginary and $q = q^*$ is real, the Bäcklund transformation (38) reduces to (29).

Belonging to Class III are also the complex MKdV equation, $q_t + 6qq^*q_x + q_{xxx} = 0$ and the Hirota equation, $q_t + 6\alpha qq^*q_x + \alpha q_{xxx} + i\beta q_{xx} + 2i\beta q^2 q^* = 0$. They all have the same spatial part of the Bäcklund transformation (38).

IV. EQUATIONS FOR HIGHER-ORDER SCATTERING PROBLEMS

We said earlier that Bäcklund transformation solutions are simple to get, especially for higher-order scattering problems. In this section, as an example, we study the Boussinesq equation describing shallow water waves

(39) $$q_{tt} = q_{xx} + (3q^2)_{xx} + q_{xxxx} .$$

Zakharov [9] proposed a set of Lax equations for it

(40)
$$4\psi_{xxx} + 3(q \frac{d}{dx} + \frac{d}{dx} q)\psi \pm \sqrt{3}\, i \int^x q_t dx\psi + \psi_x = i\lambda\psi ,$$

$$\psi_{xx} + q\psi = \pm \frac{i}{\sqrt{3}} \psi_t .$$

It is therefore a third-order scattering problem, which is difficult to solve by the inverse-scattering method. We can, however, derive a BT for it. Let $\phi \equiv \ln \psi$ and $w_x = q$, then we have

(41)
$$4(\phi_{xxx} + 3\phi_x\phi_{xx} + \phi^3) + 3(2w_x\phi_x + w_{xx}) \pm \sqrt{3}\, i\, w_t + \phi_x = i\lambda ,$$

$$\phi_{xx} + \phi_x^2 + w_x = \pm \frac{i}{\sqrt{3}} \phi_t .$$

Eliminating w, we get

(42) $$\phi_{xxx} - 2\phi_x^2 \pm 2\sqrt{3} \int^x \phi_t\phi_{xx} dx + \int^x \phi_{yy} dy + \phi_x = i\lambda .$$

The transformation $(\phi,\lambda) \to (-\phi,-\lambda)$ merely interchanges the two signs in (42). We have therefore, corresponding to the pair $(-\phi,-\lambda)$, a solution $q' = w'_x$ of (39) such that

(43)
$$4(-\phi_{xxx} + 3\phi_x\phi_{xx} - \phi_x^3) + 3(2w'_x\phi_x + w'_{xx}) \pm \sqrt{3}\, i\, w'_t - \phi_x = -i\lambda,$$

$$-\phi_{xx} + \phi_x^2 + w'_x = \mp \frac{i}{\sqrt{3}}\phi_t .$$

Now, eliminating ϕ from (41) and (43), we get the BT for the Boussinesq equation

(44)
$$\pm \frac{i}{\sqrt{3}} (w'-w)_t = (w'+w)_{xx} + (w'-w)(w'-w)_x ,$$

$$\mp \sqrt{3}\, i(w'+w)_t = (w'-w)_{xx} + (w'-w)^3 + 3(w'+w)_x(w'-w) + (w'-w)_x .$$

A superposition formula can also be derived from (44)

$$w_3 = w_1 + w_2 - w_0 + \frac{2(w_1-w_2)_x}{w_1-w_2} \quad . \tag{45}$$

Note that this formula is satisfied by solutions of the KdV equation also.

V. SUMMARY

It is demonstrated above that from the Lax equations, Bäcklund transformations between various nonlinear partial differential equations can be found. The solution of one equation implies the solution of the other. On the other hand the existence of gauge-like invariances for one equation always implies a self-Bäcklund transformation of the other. These equations having Bäcklund transformations can be grouped into classes and a part of the BT is identically the same for equations in the same class. Their solutions therefore satisfy the same superposition formula. For example, the MKdV and sine-Gordon equations have the same superposition formula

$$\tan\left(\frac{w_3 - w_0}{2}\right) = \frac{\kappa_1 + \kappa_2}{\kappa_1 - \kappa_2} \tan\left(\frac{w_1 - w_2}{2}\right) \quad . \tag{46}$$

These superposition formulas make it possible to construct N-soliton solutions using only algebraic means.

REFERENCES

[1] A.C. SCOTT, F.Y.F. CHU, AND D.W. MCLAUGHLIN, The soliton: A new concept in applied science, Proc. IEEE 61 (1973), 1443-1483.

[2] P.D. LAX, Integrals of nonlinear equations of evolution and solitary waves, Comm. Pure Appl. Math. 21 (1968), 647-690.

[3] I.M. GEL'FAND AND B.M. LEVITAN, On the determination of a differential equation from its spectral function, Amer. Math. Soc. Trans. Ser. 2 1 (1955), 253-304.

[4] C.S. GARDNER, J.M. GREENE, M.D. KRUSKAL, AND R.M. MIURA, Method for solving the Korteweg-deVries equation, Phys. Rev. Lett. 19 (1967), 1095-1097; Korteweg-deVries equation and generalizations. VI. Methods for exact solution, Comm. Pure Appl. Math. 27 (1974), 97-133.

[5] G.L. LAMB, JR., Analytical descriptions of ultrashort optical pulse propagation in a resonant medium, Rev. Modern Phys. 43 (1971), 99-124.

[6] H.D. WAHLQUIST AND F.B. ESTABROOK, Bäcklund transformation for solutions of the Korteweg-deVries equation, Phys. Rev. Lett. 31 (1973), 1386-1390.

[7] H.-H. CHEN, General derivation of Bäcklund transformations from inverse scattering problems, Phys. Rev. Lett. 33 (1974), 925-928.

[8] M.J. ABLOWITZ, D.J. KAUP, A.C. NEWELL, AND H. SEGUR, Nonlinear-evolution equations of physical significance, Phys. Rev. Lett. 31 (1973), 125-127; The inverse scattering transform - Fourier analysis for nonlinear problems, Studies in Appl. Math. 53 (1974), 249-315.

[9] V.E. ZAKHAROV, On stochastization of one-dimensional chains of nonlinear oscillators, Soviet Physics JETP 38 (1974), 108-110.

SOME COMMENTS ON BÄCKLUND TRANSFORMATIONS, CANONICAL TRANSFORMATIONS, AND THE INVERSE SCATTERING METHOD*

Hermann Flaschka and David W. McLaughlin

Department of Mathematics
The University of Arizona
Tucson, Arizona 85721

I. INTRODUCTION

In this paper we discuss two essentially independent topics. Section II is concerned with Bäcklund transformations, and is directly relevant to the main theme of this conference. Section III, on the Toda lattice equations, is connected with this main theme in a more tenuous way, only to the extent that the lattice equations are discussed through canonical theory which also is a classical transformation theory.

This paper is intended to be mainly tutorial. Much of the material will be known to some readers. However, we feel that our discussion, which takes a slightly unconventional point of view in Section II and which draws from some very recent papers in Section III, may offer some novel points to other readers. Basically, we present a variety of observations which we ourselves have found useful or interesting, but which have not been detailed in the literature. Let us list some of the main points.

Bäcklund transformations for equations such as Korteweg-deVries (KdV), sine-Gordon, etc. come in two halves--one half being particular to the specific evolution equation. In Section II we observe that the other half depends only on the associated scattering problem and can be thought of as a "change-of-potential formula" for this problem. Concentrating on this aspect of Bäcklund transformations, we show how some familiar properties (and other less familiar ones) can be derived very simply from a known formula of spectral theory. In particular we discuss the effect of Bäcklund transformations on those components of the solution (solitons and oscillations) associated with the discrete spectrum

*Work supported in part by NSF Grants GP-42739, GP-37627, and MPS 75-07530.

and continuous spectrum. Finally, we show that Bäcklund transformations are canonical transformations on the set of potentials.

Such a canonical structure plays a central role in Section III, where we discuss the inverse-scattering solution of the Toda lattice through the transformation theory of classical mechanics. Recently, this canonical interpretation has been developed in considerable detail for the partial differential equations solvable by scattering methods. On a technical level, the discussion for the lattice is quite analogous. However, it has the additional interesting feature that this description of the inverse-scattering solution for an infinitesimal motion of the Toda lattice is nothing more than the well-known action-angle solution of the harmonic lattice. This seems to be the most accessible route towards an understanding of the action-angle approach to the inverse-scattering method. Along the way to the discussion of the harmonic limit, we show how to introduce action-angle coordinates on the scattering data,[1] note the rather interesting fact that local disturbances of the Toda lattice must evolve into both solitons and continuous components, and compare this canonical method with the techniques of the Clarkson group for generating the "higher Toda lattices." Finally, we comment on the physical content of two infinite families of motion invariants and prove that the "higher polynomial constants" are equivalent to the family of action variables. The canonical notation needed to read this paper is summarized in the Appendix.

II. BÄCKLUND TRANSFORMATIONS

In this part of the paper we are concerned with Bäcklund transformations (BT's). These are usually presented as a method for transforming one solution of an evolution equation into another. They can often be derived by very classical and fairly straightforward procedures requiring lengthy differentiations and some intuition, but not much else. It is surprising that such a seemingly superficial procedure should lead, again in a fairly direct way, to the inverse-scattering

[1]Professor L.D. Faddeev has informed us that this particular computation is contained in the thesis of S.V. Manakov.

system by which various of these evolution equations can be linearized. It is even more surprising that once one has found a BT for a specific equation (say, the Korteweg-deVries equation) and thence the associated inverse problem, one is led immediately to a whole hierarchy of evolution equations, all linearizable by that same scattering problem [22]. It is probably this fact which has occasioned much of the recent interest in BT's, resulting in various attempts (reported in these proceedings) to derive BT's for specific equations by appeal to more fundamental ideas. The goal of these new approaches is a unified theory, which explains the inverse-scattering technique directly in terms of transformation structures of the evolution equation itself.

Such a comprehensive theory is yet to come. At the moment, it is the inverse method which dominates thinking about equations such as Korteweg-deVries, sine-Gordon, etc.; BT's, from this perspective, are a puzzling, if somewhat irrelevant, adjunct of the inverse method. We do not know of any problems to which BT's can be applied which cannot also be solved (and in a more systematic fashion) by the ideas of scattering theory.

As a matter of fact, very little attention has been directed towards a purely spectral-theoretic description of BT's, the main emphasis having been on transformation properties of specific evolution equations. Here we shall largely ignore the latter approach, and view BT's as a transformation of scattering problems. It so happens that certain of the formulas discussed at this conference are already familiar in a different context: the spectral theory of Sturm-Liouville problems. Simple arguments based on these formulas allow us to describe clearly and concisely the possible influence a BT may have on the soliton and the continuous-spectrum (ringing) parts of a solution of a nonlinear equation. Entirely analogous arguments apply as well when periodic boundary conditions are imposed, e.g. on the KdV equation, and these lead to qualitative versions of the results of Hirota [16] and Wahlquist [31] on the asymptotic phase shift of cnoidal waves perturbed by a soliton. It will be seen that the spectral properties of BT's are extremely trivial, a fact which in our opinion only reinforces the need for a more thorough understanding of these simple transformations which manage (by their mere existence) to indicate that an evolution equation has interesting properties.

It is time to be more specific. To illustrate the phenomena alluded to above, we recall certain facts [32] about the BT for the KdV equation, $q_t - 6qq_x + q_{xxx} = 0$. In terms of w, defined through $w_x \equiv q/2$, a new solution $Q \equiv 2W_x$ of the KdV equation is defined through

$$W_x = -w_x + \lambda + (W - w)^2 , \tag{1}$$

$$W_t = -w_t + 4[\lambda W_x + w_x^2 + w_x(W - w)^2 + w_{xx}(W - w)] . \tag{2}$$

Setting $W - w = -y_x/y$, we find for y the equation

$$-y_{xx} + qy = \lambda y . \tag{3}$$

This, of course, is the eigenvalue problem on which the complete solution of KdV can be based [14], [15]. This same eigenvalue problem can, however, be used as well to solve the trivial equation $q_t = q_x$, a certain fifth-order equation $q_t + q_{xxxxx} + \ldots = 0$, and in fact infinitely many others [22], [33]. Conversely, one might expect that all these equations possess BT's, and that all these BT's would contain (1), since it is that relation which gives rise to the eigenvalue problem (3). We have already indicated that we wish to concentrate on the latter; accordingly, we shall view (1) as a transformation of coefficients in (3). Thus, we think of $\ell \equiv -\frac{d^2}{dx^2} + q$ as the "old" operator, and of $L \equiv -\frac{d^2}{dx^2} + Q$ as the "new" one, and we ask how the spectral data of the new operator differ from those of the old one.

1. The Basic Formula

Most of our deductions will follow from one (known) formula which shows how eigenfunctions of L may be expressed in terms of eigenfunctions of ℓ [7].[2]

[2] The relevance of this formula for BT's was noted independently in [30], and used there to generate the BT for KdV (in the absence of continuous spectrum).

Let y_0 be a fixed solution of $\ell y = \lambda_0 y$, and let $y(x,\lambda)$ satisfy $\ell y = \lambda y$. Then

$$Y(x,\lambda) = y'(x,\lambda) - y(x,\lambda)\,\frac{y_0'(x,\lambda_0)}{y_0(x,\lambda_0)}$$

solves $LY = \lambda Y$. If $\lambda = \lambda_0$, two independent solutions of $Ly = \lambda_0 y$ are

$$\frac{1}{y_0(x,\lambda_0)} \quad \text{and} \quad \frac{1}{y_0(x,\lambda_0)} \int^x y_0^2(z,\lambda_0)\,dz \ .$$

Observe that these formulas are meaningless if ever $y_0 = 0$; according to the oscillation theorems, the transforming parameter λ_0 should lie to the left of all continuous or discrete eigenvalues of ℓ if this singularity is to be avoided (λ_0 is allowed to equal the smallest eigenvalue, but for simplicity, we shall assume it to lie strictly to the left of the spectrum in what follows).

2. Transformation of Scattering Data

In this subsection, we assume that $q(x)$ decreases rapidly as $x \to \pm\infty$. Then one can identify, for any k with $\operatorname{Im} k \geq 0$, certain solutions of $\ell y = k^2 y$ by their asymptotic behavior:

$$f_1(x,k) \sim \begin{cases} e^{ikx}, & x \to +\infty, \\ b(k)e^{-ikx} + a(k)e^{ikx}, & x \to -\infty, \end{cases}$$

(4)

$$f_2(x,k) \sim \begin{cases} -b^*(k)e^{ikx} + a(k)e^{-ikx}, & x \to +\infty, \\ e^{-ikx}, & x \to -\infty. \end{cases}$$

According to inverse-scattering theory, the coefficient q can be reconstructed from the function $b(k)/a(k)$, the proper eigenvalues (if present) $-\eta_1^2,\ldots,-\eta_N^2$, and the normalization constants $c_1,\ldots,c_N$ defined in terms of the asymptotic behavior $y_j \sim \sqrt{c_j}\,\exp(-\eta_j x)$ of the normalized eigenfunctions, i.e. $\int_{-\infty}^{\infty} y_j^2(x)\,dx = 1$. It is easy to compute the effect of the transformation $\ell \to L$ on these scattering data. We write A, B, C_j for the transforms of a, b, c_j.

Proposition: Fix $\lambda_0 = -\eta^2$ (restricted as described earlier). Let y_0 be a solution of $\ell y = \lambda_0 y$.

i) If $y_0 = f_1(x,i\eta)$, then

$$A(k) = a(k), \quad B(k) = \frac{\eta - ik}{\eta + ik} b(k) ,$$

the eigenvalues of ℓ and L are identical, and

$$C_j = \frac{\eta - \eta_j}{\eta + \eta_j} c_j .$$

ii) If $y_0 = f_2(x,i\eta)$, then

$$A(k) = a(k), \quad B(k) = \frac{\eta + ik}{\eta - ik} b(k) ,$$

the eigenvalues of ℓ and L are identical, and

$$C_j = \frac{\eta + \eta_j}{\eta - \eta_j} c_j .$$

iii) If $y_0 = D_1 f_1(x,i\eta) + D_2 f_2(x,i\eta)$, D_1, D_2 constants, then

$$A(k) = \frac{ik + \eta}{ik - \eta} a(k), \quad B(k) = -b(k) ,$$

L has all eigenvalues of ℓ plus the new eigenvalue $-\eta^2$, with normalization constant depending on D_1 and D_2, while the normalization constants corresponding to the old eigenvalues transform according to

$$C_j = \frac{\eta + \eta_j}{\eta - \eta_j} c_j .$$

Proof: To prove the various assertions about $A(k)$ and $B(k)$, one must only use the basic formula in Subsection 1 to find the solution $F_1(x,k)$ of $Ly = k^2 y$ which goes like $\exp(ikx)$ at $+\infty$. The new scattering data are then read off at $-\infty$ according to the asymptotic behavior $F_1(x,k) \sim B(k)e^{-ikx} + A(k)e^{ikx}$. In case i), for example, one takes $y(x,\lambda) = f_1(x,k)/(\eta+ik)$; it is then routine to check that the corresponding new solution $Y(x,\lambda)$ is in fact $F_1(x,k)$. These manipulations, and the ones needed in the determination of the C_j, are straightforward and are left to the reader.

Below we state a number of observations on the Proposition.

Observation 1. There are BT's which do not add eigenvalues (cases i, ii). In more picturesque terms, they do not add solitons, but they do shift any solitons that might be present in the old coefficient q (since $C_j \neq c_j$, cf. [15]), and they do change the phase of the continuous spectrum component by a factor $\frac{\eta \pm ik}{\eta \mp ik}$ of modulus 1.

Observation 2. Those BT's (case iii) which add a new eigenvalue (or soliton) shift the phase of the continuous spectrum by $e^{i\pi}$ (since $b(k) \to -b(k)$).

Observation 3. Let us call factors such as $\frac{\eta - ik}{\eta + ik}$, etc., "transformation factors." It is clear that composition of BT's corresponds to multiplication of the transformation factors; this relationship is reminiscent of the theorem about Fourier transforms of convolutions. Indeed, in the limit "q is infinitesimal," the transform to scattering data becomes essentially a Fourier transform (see Section III for more details), and the BT does become a convolution with a certain kernel. We do not present the explicit formulas, since their significance escapes us at present.

Several examples of this multiplicative property are noteworthy.

Example: Transformations of types i and ii with the same λ_0 are inverses of each other. All the scattering data are restored to their original values, e.g. $b \to \frac{\eta - ik}{\eta + ik} b \to \frac{\eta + ik}{\eta - ik} \frac{\eta - ik}{\eta + ik} b$ so that the initial coefficient $q(x)$ must agree with the final one.

Example: The transformation

$$q \xrightarrow{f_2(x,i\eta)} Q \xrightarrow{D_1F_1(x,i\eta) + D_2F_2(x,i\eta)} \tilde{Q}$$

adds one eigenvalue, $-\eta^2$, but leaves the reflection coefficient $b(k)/a(k)$ and the original normalization constants c_j unchanged. The final coefficient is given by

$$\tilde{q}(x) = q(x) - 2\left[\frac{D\, f_2^2(x,i\eta)}{1 + D\int_{-\infty}^{x} f_2^2(z,i\eta)\,dz}\right]', \qquad D = \text{constant}$$

This is an interesting example, because the familiar facts about BT's adding on solitons might lead one to conjecture that a BT should be defined as a transformation which adds one eigenvalue and leaves all other spectral data unchanged. We see here that this requirement is implemented by <u>two</u> separate BT's. Only if the initial $q(x)$ is zero does one of these two transformations reduce to the identity. The reader familiar with the inverse-scattering method will be able to verify quite easily that the solution of the Gel'fand-Levitan-Marchenko equation describing the addition of exactly one eigenvalue leads to the formula just given for $\tilde{q}$.

We think this perhaps the most curious of our observations: that the usual BT, applied to a general coefficient $q(x)$, does have a nontrivial effect on the continuous spectrum.

<u>Observation 4</u>. It might be interesting to characterize that class of coefficients which can be obtained from a given q by application of a sequence of BT's. For example, if no eigenvalues are added, one would generate a transformation of $b(k)$ of the type

$$b(k) \to \prod_j \frac{\alpha_j \mp ik}{\alpha_j \pm ik}\, b(k)$$

The ratio: new b/old b would thus be of a very special type. In studying this question, one would apparently be led to consider scattering matrices differing by "inner factors" or, more specially, Blaschke products.

<u>Observation 5</u>: To this point, we have avoided all mention of time-evolution; our discussion was confined to the first of the two BT equations (1). It is a simple matter, however, to incorporate time-evolution in the spectral description of BT's. For example, let $q(x,t)$ be a solution of KdV. Then we know that the eigenvalues of $-\frac{d^2}{dx^2} + q(x,t)$, as well as the scattering function $a(k,t)$, are independent of time, while

$$(5) \qquad \frac{d}{dt} b(k,t) = 8ik^3 b(k,t), \ \frac{d}{dt} c_j(t) = 8\eta_j^3 c_j(t).$$

Imagine performing a BT, i.e. a transformation of coefficients $q(x,t) \to Q(x,t)$ as envisaged in the Proposition, at each instant t, always using the function $f_1(x,i\eta;t)$ (which now depends on t). Then $B(k,t) = \frac{\eta - ik}{\eta + ik} b(k,t)$. Hence, $B(k,t)$ satisfies the first equation in (5), and likewise, $C_j(t)$ will satisfy the second. It follows that $Q(x,t)$ again solves KdV. One might say that the BT commutes with the KdV flow. In principle, the standard form of the BT (1), (2), can be derived from the facts just described; in practice, the computations are horrible.

This concludes our observations on the Proposition. Let us summarize the main points:

The usual BT for KdV can be described entirely in terms of the associated linear problem, as a certain kind of transformation of scattering data. The latter is quite trivial, consisting of a change of phase in the scattering data, and/or the addition of one eigenvalue.

This very simplicity of BT's (as spectral transformations) poses new questions. It is plausible that BT's, viewed as transformations of an evolution equation, may have a real significance. Transformation properties of differential equations generally lie at the heart of special solution methods, and the intricate structures revealed by Wahlquist and Estabrook [32] in their study of KdV certainly suggest that the inverse-scattering method is no exception. It is puzzling that the spectral description of BT's should be so uncomplicated. One suspects that the kind of transformations described in the Proposition should play a role of some importance in the scattering problem, but as yet we have no ideas about their significance.

We might note that similar computations could be performed in connection with the known BT's for certain other evolution equations. The modified Korteweg-deVries (MKdV) equation, $q_t + 6q^2 q_x + q_{xxx} = 0$, can be solved by the scattering problem for the system [1], [29].

$$v_{1x} + i\zeta v_1 = qv_2 ,$$
$$v_{2x} - i\zeta v_2 = -qv_1 .$$

The analog of (1) is known to be [1], [30].

$$Q = q - 2(\tan^{-1}\Gamma)_x$$

where $\Gamma = \dfrac{v_1(x_1,-i\eta_0)}{v_2(x_1,-i\eta_0)}$. It can be shown that new eigenfunctions are related to the old ones according to the rule[3]

$$(6) \qquad \begin{pmatrix} V_1 \\ V_2 \end{pmatrix} = \begin{pmatrix} v_1 \\ v_2 \end{pmatrix}_x + \eta_0 \begin{pmatrix} v_1 \\ v_2 \end{pmatrix} - \frac{2\eta_0}{1+\Gamma^2} \begin{pmatrix} v_1 \\ v_2 \end{pmatrix} + \begin{pmatrix} 0 & \dfrac{-\Gamma_x}{1+\Gamma^2} \\ \dfrac{\Gamma_x}{1+\Gamma^2} & 0 \end{pmatrix} \begin{pmatrix} v_1 \\ v_2 \end{pmatrix} .$$

One can also recast the linear problem associated with KdV as a first-order system, namely

$$v_{1x} + i\zeta v_1 = qv_2 ,$$
$$v_{2x} - i\zeta v_2 = v_1 .$$

One finds then that $Q = q - 2\Gamma_x$, while the formula of Subsection 1 assumes the form

$$(7) \qquad \begin{pmatrix} V_1 \\ V_2 \end{pmatrix} = \begin{pmatrix} v_1 \\ v_2 \end{pmatrix}_x + \eta_0 \begin{pmatrix} v_1 \\ v_2 \end{pmatrix} - \Gamma \begin{pmatrix} v_1 \\ v_2 \end{pmatrix} + \begin{pmatrix} 0 & -\Gamma_x \\ 0 & 0 \end{pmatrix} \begin{pmatrix} v_1 \\ v_2 \end{pmatrix} .$$

Clearly there are similarities between (6) and (7).

3. Periodic Problems

It has on occasion been suggested that BT techniques might enable one to construct interacting cnoidal-wave solutions of the periodic KdV. The recent solutions, by various authors, [6], [18], [24] of the periodic problem have revealed the very complex structure of this problem; in retrospect one would be very surprised had a single, direct technique like the BT led to a resolution of this problem. The explicit computations of Wahlquist [31] and Hirota [16] had already

[3] A similar equation probably holds for the system associated with the nonlinear Schrödinger equation.

indicated that the BT of a cnoidal-wave is still only a cnoidal-wave, shifted by a certain constant. These authors further observed that one could construct a solution representing a soliton riding over a cnoidal-wave, and they observed that there was an asymptotic phase shift in the underlying periodic wave train. This is especially clear in the elegant formulas of Wahlquist. We shall now explain how this type of information can be deduced from the formula of Subsection 1. While this approach does not require restriction to cnoidal solutions, its results are necessarily of a purely qualitative nature.

We will show the following: If one requires the new potential to be periodic, one must perform BT's which are the analogs of cases i) and ii) in the Proposition. These do not increase the "degrees of freedom" in the infinite case, and should not be expected to do so in the periodic problem. If, as in case iii) above, one adds a soliton, one can read off certain asymptotic facts from the transformation formulas.

We recall that when q is periodic (say of period 1), the spectral theory of $\ell \equiv -\frac{d^2}{dx^2} + q$ is best approached by a study of the so-called Floquet solutions y_+, y_- of $\ell y = \lambda y$. These satisfy, respectively,

$$y_+(x+1) = \rho y_+(x), \quad y_-(x+1) = \rho^{-1} y_-(x) .$$

The multiplier ρ is a function of λ. Since two solutions of this type form a basis for all solutions of $\ell y = \lambda y$, it is seen that any solution of this equation is bounded in x if $|\rho| < 1$ and is unbounded at $+\infty$ or $-\infty$ if $|\rho| > 1$. It is known that the latter holds for all sufficiently negative λ. The set of λ's for which solutions are bounded has the interval-gap structure familiar from solid-state physics or stability theory, i.e. this set is the union of infinitely many intervals which extend to $+\infty$ and are, in general, separated by "gaps" in each of which $|\rho(\lambda)| > 1$.

It may happen, for special choices of q, that all but a finite number of these gaps disappear, so that from some λ_1 onward, all solutions of $\ell y = \lambda y$ turn out to be bounded. It has recently become clear that the correct periodic analog of an N-soliton potential is precisely a $q(x)$ for which all but N of these spectral gaps disappear.[4] If there is only one gap, a theorem of Hochstadt [17] asserts that q must be a cnoidal function; this is the spectral characterization of the cnoidal-wave solution of KdV. The "interaction of two cnoidal wave-trains" referred to in various earlier papers must apparently be understood to be a KdV solution whose Schrödinger operator has precisely two gaps.

It is evident from this discussion, that the question "Can BT produce a KdV solution with more degrees of freedom?" has the spectral counterpart: "Can BT increase the number of gaps in the spectrum of a periodic Schrodinger operator?" It is easy to see what the answer must be: no.

If the new potential $Q = q - 2(\ln y_0)''$ is to be periodic, then y_0'/y_0 must be periodic, and this can only happen if y_0 is one of the two Floquet solutions, $y_{\pm}$. For a given λ_0, then, there are two, and only two, BT's that yield new periodic potentials. They are the counterparts of cases i) and ii) in the Proposition. It is now trivial to check, using Subsection 1, that a new eigenfunction $Y(x,\lambda)$ is bounded if and only if the old one, $y(x,\lambda)$, was bounded. Since a spectral gap consists of those λ's for which solutions of $LY = \lambda Y$ are unbounded, one concludes that no new spectral gaps could have been opened up by this kind of BT. If, in particular, the original q is cnoidal, i.e. has one spectral gap, then the new Q still has the same gap; by the aforementioned result of Hochstadt, Q must again be cnoidal (with the same speed, as this is determined by the gap size). This explains the result of Wahlquist and Hirota.

Suppose one wants to add a soliton; then one must add an eigenvalue to the otherwise continuous spectrum of ℓ. This is done as in case iii) of the Proposition. One lets $y_0 = D_1 y_+ + D_2 y_-$. It is now easy to see, by checking the asymptotic behavior of $Q = q - 2(\ln y_0)''$, that Q at $+\infty$ behaves as though only y_+ had been used in the BT, while Q at $-\infty$ behaves as though only y_- had been used. Thus, the underlying periodic wave suffers asymptotic distortions,

[4] For a simple-minded discussion of the evidence, see [10]. The complete theory of N-gap potentials has recently been worked out by a number of authors [6], [18], [24].

which are described precisely by the two periodic BT's. This distortion for a cnoidal q can only be a phase shift; for more complicated q it will, in general, be a change of shape as well.

It is apparent that the addition of an eigenvalue entails a non-local perturbation of the original periodic potential. One can make the perturbation local at either $+\infty$ or $-\infty$ (but not both) by superposing two BT's. Let λ_0 lie in a spectral gap. Then y_0'/y_0 must have zeros, and will be singular. One may, however, proceed in a purely formal way, choosing $y_0 = y_-$, and then performing a second BT with the new eigenfunction $Y_0 = [c \int_x^\infty y_0^2 + 1]/y_0$. The composition of these two transformations is nonsingular, being given by

$$\mathfrak{q} = q - 2(\ln[1 + c \int_x^\infty y_0^2])''.$$

One can see that $\mathfrak{q} \sim q$ at $+\infty$, while $\mathfrak{q}$ and q are different at $-\infty$.

It is a remarkable fact that a truly local perturbation (e.g. compact support) must change the spectrum drastically. It has been shown by Zheludev [35], that a local perturbation of constant sign <u>must</u> introduce a proper eigenvalue (with square-integrable eigenfunction) in every spectral gap! Presumably, this has implications about solutions of the KdV equation. It suggests that such perturbations must break up into infinitely many solitons riding over a periodic background wave. The new eigenvalues, by the way, must increase like n^2 so it is doubtful whether the familiar relation between magnitude of eigenvalue and speed of soliton will remain in force. The general picture is not entirely implausible, since it is known from earlier perturbation arguments that solitary waves over an uneven bottom must break up; this appears to be the fate of the initial local disturbance in the present example as well.

4. Canonical Structure[5]

The connection between the KdV equation and the Schrödinger operator has become more than a means for solving KdV--it has stimulated an extensive and deep

[5] Some familiarity with canonical theory, as described in the Appendix, is assumed in this section.

re-examination of the spectral properties of $-\frac{d^2}{dx^2} + q(x)$, particularly with periodic q. One important new discovery is the following: There is a natural Poisson bracket defined on functionals of the potential q,

$$\{F(q),G(q)\} \equiv \frac{1}{2}\int_0^1 \left(\frac{\delta F}{\delta q}\frac{d}{dx}\frac{\delta G}{\delta q} - \frac{\delta G}{\delta q}\frac{d}{dx}\frac{\delta F}{\delta q}\right) dx, \tag{8}$$

($\frac{\delta}{\delta q}$ denotes the functional derivative). Any functional of q may serve as "Hamiltonian," and the Poisson bracket provides a prescription for finding the corresponding Hamiltonian equations [13]. It is easy to check that these take the form

$$q_t = \frac{d}{dx}\frac{\delta F(q)}{\delta q} \quad .$$

(Compare (A.2); the skew-symmetric operator C is $\frac{d}{dx}$, which is not invertible). Of particular importance are the Hamiltonian equations generated by the constants of motion of the KdV equation. These constants (thought of as functionals of q) are all in involution, and the corresponding equations are precisely: $q_t = q_x$; the KdV itself; and the higher-order KdV-like equations found by Lax [22] and Lenard (see [15]). A more detailed exposition of the canonical structure associated with an eigenvalue problem will be found in Section III. That discussion (and numerous other recent papers) make it plain that canonical structures, such as the one defined by (8), are an essential feature of the inverse method. It is then quite natural to inquire about the role of BT's in this canonical theory, and the answer is simple:

<u>The BT $q \to Q$ is a canonical transformation (with respect to the Poisson bracket (8)) on the set of periodic potentials constrained by</u>[6] $\int_0^1 q(x)dx =$ <u>constant</u>.

It should be emphasized that we are again concerned with only the first half of the KdV BT (1). The concept of "canonical transformation" might also arise in a more classical setting, in connection with the transformation properties of

[6] This restriction arises from the fact that a constant $q(x)$ lies in the null space of the skew-symmetric operator d/dx. Compare also with Remark 4 in Section III.

the KdV equation after the latter is written as a system of differential forms. This is decidedly different from our use of the term.

To prove the result, one must verify the invariance of the Poisson bracket under the transformation $q \to Q$. Alternatively, one may make use of the fact that the Poisson bracket (8) is associated with a certain two-form on the set of potentials, and verify the invariance of this form. The details are lengthy, and not in themselves very instructive, and so are omitted.[7]

The implications of this result do deserve some further comment. It is interesting, first of all, that the BT, viewed as transformation of potentials, should be of a classical type (if in a non-classical setting); this fact stands in sharp contrast to the relative novelty of the complete BT (1), (2), as transformation of a partial differential equation. Secondly, it is a basic fact of canonical theory that canonical transformations preserve the form of Hamilton's equations; this provides a deeper explanation of the fact alluded to in Observation 5 above: that BT commutes with the KdV flow. Finally, one knows that many canonical transformations have generating functions, and it is natural to ask whether this is true of the BT. We have not yet found a generating functional; one might hope that if one exists, it will turn out to be recognizable as a quantity of some importance in spectral theory. In that way, a better

---

[7] The second method seems somewhat more direct, and is the one used in our proof. Let us describe briefly what must be proved. The two-form ω is defined on vectors tangent to the set of potentials, i.e. let $q(x;\epsilon)$ be a one-parameter family of potentials, and expand near $\epsilon = 0$: $q(x;\epsilon) = q(x;0) + \epsilon\xi(x) + o(\epsilon)$; the function $\xi(x)$ represents a tangent vector at $q(x,0)$. If ξ and η are tangent vectors at q, one defines

$$\omega(\eta,\xi) = \int_0^1 \eta(x)\ (A\xi)(x)dx, \quad \text{where } A = \frac{1}{4}\left(\int_x^\pi - \int_0^x\right).$$

Now let $q(x,\epsilon)$ be a curve through $q(x)$, and let $Q(x;\epsilon)$ be the curve through $Q(x)$ obtained by performing (for each ϵ) a BT with fixed parameter λ_0 and eigenfunction $y_0(x,\lambda_0;\epsilon)$. The tangent vector $\xi(x)$ at q transforms into a tangent vector $\Xi(x)$ at Q; the formula of Subsection 1 enables one to find the transformation explicitly. The verification of the invariance of the two-form,

$$\omega(\eta,\xi) = \omega(H,\Xi)$$

then becomes a lengthy exercise in integration by parts.

understanding of the spectral theory of BT's may follow.

III. THE TODA LATTICE

1. Description of the Toda Lattice

The Toda lattice [27], [28] is a chain of unit masses connected by nonlinear springs subject to longitudinal vibrations. The potential energy function is

$$V(r) = e^{-r} + r - 1 ,$$

r being the elongation of the spring. The equations of motion for this lattice are derived from the Hamiltonian

$$H = \sum_{n=-\infty}^{\infty} \frac{1}{2} P_n^2 + V(Q_n - Q_{n-1}) ,$$

in which Q_n is the displacement from equilibrium of the nth mass[8] and P_n its momentum. For convenience, we introduce the ∞-dimensional vectors $Q = (\cdots,Q_{-1}, Q_0,Q_1,\cdots)$ and $P = (\cdots,P_{-1},P_0,P_1,\cdots)$, and write the Toda equations in block matrix form

$$\begin{pmatrix} \dot{Q} \\ \dot{P} \end{pmatrix} = \begin{pmatrix} 0 & I \\ -I & 0 \end{pmatrix} \begin{pmatrix} \nabla_Q H \\ \nabla_P H \end{pmatrix} \equiv T \ \mathrm{grad}_{(Q,P)} H ,$$

with

$$H \equiv \frac{1}{2} P^2 + V(Q), \quad T \equiv \begin{pmatrix} 0 & I \\ -I & 0 \end{pmatrix} .$$

We shall have occasion to compare this nonlinear lattice with the <u>harmonic</u> one, whose potential is $\tilde{V}(r) = \frac{1}{2} r^2$. The abbreviated notation will be used in this case as well.

One can sum the linear term in the Toda potential to obtain

$$V(Q) = \sum_{n=-\infty}^{\infty} [\exp(Q_{n-1} - Q_n) - 1] + (Q_\infty - Q_{-\infty}) .$$

---

[8] The actual location of the nth particle is given by $n\ell + Q_n$, where ℓ denotes the separation of mass points at equilibrium.

The term $(Q_{\infty} - Q_{-\infty})$ is the total elongation of the Toda lattice (from its equilibrium length); as we shall explain below, it will be nonzero in some important cases, and so must be retained.

The usual way of solving the harmonic lattice problem relies on Fourier transform methods. The explicit transformations are summarized in Table 1. In the physical variables $(\tilde{Q},\tilde{P})$, the equations, while linear, are coupled. Fourier methods not only decouple the system, they also lead to a very special set of canonically conjugate variables--the <u>normal modes</u>. In this normal mode representation, the decoupled dynamics depends explicitly on the <u>dispersion relation</u> $\omega(k)$. One may either integrate the equations in this form or perform an additional canonical transformation which leads to <u>action-angle</u> coordinates. In this latter representation, the equations remain decoupled, the dynamics still depends explicitly on the dispersion relation $\omega(k)$, and half the canonical variables (the action variables, $\tilde{J}(k)$) are <u>constants of the motion</u>. The Toda lattice problem can be solved in much the same way through the inverse scattering transform (IST).

2. Canonical Description of the IST Solution of the Toda Lattice

Flaschka [8], [9] (see also [23]) solved the Toda lattice equations via scattering theory by first transforming from the physical (Q,P) variables to new ones,

$$a_n = \frac{1}{2} \exp[(Q_{n-1} - Q_n)/2], \quad b_n = -\frac{1}{2} P_{n-1} .$$

Under the restriction $Q_n \to 0$ as $n \to -\infty$, the transformation has a unique inverse.

We shall use the symbol S to denote the collection of sequences (Q,P) satisfying: (i) $Q_n \to 0$ as $n \to -\infty$, (ii) $(Q_n - Q_{n-1}) \to 0$ as $n \to \pm\infty$, and (iii) $P_n \to 0$ as $n \to \pm\infty$; and the symbol $\bar{S}$ to denote the collection of sequences (a,b) satisfying: (i) $a_n \geq 0$, (ii) $a_n \to \frac{1}{2}$ as $n \to \pm\infty$, and (iii) $b_n \to 0$ as $n \to \pm\infty$. Each of these limits is supposed to be approached sufficiently rapidly for our manipulations to be valid. We will not specify the precise rates of convergence.

Harmonic Lattice

1. **Physical:** $(\tilde{\mathcal{Q}},\tilde{\mathcal{P}})$

$$\tilde{H} = \sum_n \frac{1}{2}\left[\tilde{P}_n^2 + (\tilde{Q}_n - \tilde{Q}_{n-1})^2\right] .$$

2. **Normal Modes (a):** $(\tilde{\mathcal{Q}},\tilde{\mathcal{P}})$

$$\tilde{\mathcal{Q}}(k) = (2\pi)^{-\frac{1}{2}}\sum_j Q_j e^{-ikj}$$

$$\tilde{\mathcal{P}}(k) = (2\pi)^{-\frac{1}{2}}\sum_j P_j e^{ikj} , \; k \in (-\pi,\pi) .$$

$$\tilde{H} = \int_{-\pi}^{\pi} dk \, \frac{1}{2}\left[\tilde{\mathcal{P}}(k)\tilde{\mathcal{P}}(-k) + \omega^2(k)\tilde{\mathcal{Q}}(k)\tilde{\mathcal{Q}}(-k)\right] .$$

3. **Normal Modes (b):** $(\tilde{A}_+,\tilde{A}_-)$

$$\tilde{A}_+(k) = \left[|\omega(k)|\tilde{\mathcal{Q}}(k) - i\tilde{\mathcal{P}}(-k)\right]/\sqrt{2|\omega(k)|}$$

$$\tilde{A}_-(k) = -i\left[|\omega(k)|\tilde{\mathcal{Q}}(-k) + i\tilde{\mathcal{P}}(k)\right]/\sqrt{2|\omega(k)|} ,$$

$$k \in (-\pi,\pi) .$$

$$\tilde{H} = \int_{-\pi}^{\pi} dk \; i|\omega(k)|\left[\tilde{A}_+(k)\tilde{A}_-(k)\right] .$$

4. **Action-Angle:** $(\tilde{\theta},\tilde{J})$

$$\tilde{\theta}(k) = \frac{1}{2i}\,\ln\left[\frac{\omega(k)}{i}\frac{\tilde{\mathcal{Q}}(k)}{\tilde{P}(k)}\right] .$$

$$\tilde{J}(k) = \left[\tilde{\mathcal{P}}(k)\tilde{P}(-k) + \omega^2(k)\tilde{\mathcal{Q}}(k)\tilde{\mathcal{Q}}(-k)\right]/2\omega(k) ,$$

$$k \in (-\pi,\pi) .$$

$$\tilde{H} = \int_{-\pi}^{\pi} dk \; |\omega(k)| \; \tilde{J}(k) .$$

Toda Lattice

1. **Physical:** (Q,P)

$$H = \sum_n \frac{1}{2}P_n^2 + \left[e^{(Q_{n-1}-Q_n)} - 1 + (Q_{n-1}-Q_n)\right] .$$

2. **Normal Modes:** (A_+,A_-)

$$H = \int_0^{\pi} dk \, i\omega(2k)A_+(k)A_-(k) + \sum_{j=1}^{N}\left[\frac{1}{2}\left(z_j^{-2}-z_j^2\right) + \ln z_j^2\right]$$

3. **Action-Angle:** (θ,J)

$$H = \int_0^{\pi} dk \; \omega(2k)J(k) + \sum_{j=1}^{N}\left[\frac{1}{2}\left(z_j^{-2} - z_j^2\right) + \ln z_j^2\right]$$

Table 1. Canonical variables $[\omega(k) = 2 \sin \frac{1}{2}(k)]$.

We begin to investigate the canonical nature of the map from the (Q,P) variables to the (a,b) variables by computing Poisson brackets among a_n and b_m taken as functions of Q and P:

$$\{a_n,b_m\}_{\mathcal{T}} \equiv \sum_{\ell}\left(\frac{\delta a_n}{\delta Q_\ell}\frac{\delta b_m}{\delta P_\ell} - \frac{\delta a_n}{\delta P_\ell}\frac{\delta b_m}{\delta Q_\ell}\right) = \frac{1}{4}\, a_n(\delta_{n+1,m} - \delta_{n,m}) ,$$

$$\{a_n,a_m\}_{\mathcal{T}} = \{b_n,b_m\}_{\mathcal{T}} = 0 .$$

Here $\mathcal{T} \equiv \begin{pmatrix} 0 & I \\ -I & 0 \end{pmatrix}$ and the notation is discussed in the Appendix. Now define the skew-symmetric block matrix $\mathcal{C}$ by

$$\mathcal{C} \equiv \frac{1}{4}\begin{pmatrix} 0 & C \\ -C^T & 0 \end{pmatrix}, \quad C_{nm} \equiv a_n(\delta_{n+1,m} - \delta_{n,m}) ,$$

and use it to define Poisson brackets for functions of a and b according to the Appendix. We find, for instance, that

$$\{a_n,b_m\}_{\mathcal{C}} \equiv \left(\mathrm{grad}_{(a,b)}a_n,\ \mathcal{C}\ \mathrm{grad}_{(a,b)}b_m\right)$$

$$= \begin{pmatrix} \nabla_a a_n \\ \nabla_b a_n \end{pmatrix} \cdot \begin{pmatrix} 0 & \frac{1}{4}C \\ -\frac{1}{4}C^T & 0 \end{pmatrix} \begin{pmatrix} \nabla_a b_m \\ \nabla_b b_m \end{pmatrix}$$

$$= \frac{1}{4}\sum_{\ell,\ell'} \delta_{n\ell} C_{\ell\ell'} \delta_{\ell' m} = \frac{1}{4} C_{nm} = \{a_n, b_m\}_{\mathcal{T}}.$$

In this way one establishes the relations

$$\{a_n,a_m\}_{\mathcal{T}}=\{a_n,a_m\}_{\mathcal{C}}=0; \quad \{b_n,b_m\}_{\mathcal{T}}\{b_n,b_m\}_{\mathcal{C}}=0; \quad \{a_n,b_m\}_{\mathcal{T}}=\{a_n,b_m\}_{\mathcal{C}} .$$

This invariance of the Poisson brackets guarantees that the map from (Q,P) to (a,b) is a canonical transformation between $\mathcal{S}$ (coordinatized by (Q,P) with Poisson bracket characterized by $\mathcal{T}$) and $\bar{\mathcal{S}}$ (coordinatized by (a,b) with Poisson bracket characterized by $\mathcal{C}$).

Notice that the null space of $\mathcal{C}$ is two-dimensional; it is spanned by

$$\begin{pmatrix}0\\1\end{pmatrix} \text{ and } \begin{pmatrix}\frac{1}{a}\\0\end{pmatrix}, \text{ where } \left(\frac{1}{a}\right)(n) \equiv \frac{1}{a_n}, \quad (1)(n) \equiv 1 .$$

Thus, C is singular (see Appendix). The null space of C will show up again in the dynamics of the Toda lattice.

Since the map from the (Q,P) description of the lattice to the (a,b) description is canonical, we immediately know that the Toda lattice equations transform to

$$\begin{pmatrix}\dot{a}\\b\end{pmatrix} = C \operatorname{grad}_{(a,b)} H$$

where

$$H[a,b] \equiv H[Q(a,b),P(a,b)] = \sum_n [(4a_n^2 - 1) + 2b_n^2] + 2 \ln(\Pi 2a_n)^{-1} .$$

Explicitly,

$$\dot{a}_n = a_n(b_{n+1} - b_n) ,$$

$$\dot{b}_n = 2(a_n^2 - a_{n-1}^2) ,$$

if one uses the fact that the sequence $\{a_n^{-1}\}$ belongs to the null space of C.

The initial-value problem for the Toda lattice in $\bar{S}$ is solved by first mapping the initial data $(a(t=0), b(t=0))$ into certain <u>scattering data</u> $\Sigma(t=0)$ (reflection coefficients, point spectrum, and normalization constants). The temporal evolution of the scattering data $\Sigma(t)$, being linear and decoupled, is explicitly calculated. Then $(a(t), b(t))$ are constructed from $\Sigma(t)$ by a discrete inverse scattering theory. The details of this inverse scattering solution of the Toda lattice have been clearly discussed elsewhere [9], [23], [28]. Here we merely summarize the explicit maps in Table 2, and concentrate on (i) the canonical nature of the map to scattering data and (ii) the relationship of this transformation to the normal mode solution of the harmonic lattice.[9]

---

[9]Since this paper was completed, we have shown these continuous action variables to be limits of action variables defined, in the classical manner, as loop integrals of the type $\oint pdq$ over cycles in the phase space of the periodic Toda lattice. See [11].

1.	Linear Problem:	$L\psi = \lambda\psi$, $(L\psi)(n) = a_{n-1}\psi_{n-1} + a_n\psi_{n+1} + b_n\psi_n$.
2.	Spectrum of L:	Point Spectrum $z_j \in (-1,1)$ for $j = 1,\dots,N$, Continuous Spectrum $z = e^{ik}$ for all $k \in [0,2\pi]$, $[\lambda = \frac{1}{2}(z + z^{-1})]$
3.	Boundary Conditions:	"Jost Solutions" $\phi(n,z) \underset{\sim}{} z^n$ as $n \to +\infty$, $\psi(n,z) \underset{\sim}{} z^{-n}$ as $n \to -\infty$, "Bound States" $\sum_{n=-\infty}^{\infty} \zeta^2(n,z_j) = 1$ for $j = 1,\dots,N$.
4.	Relationships:	$\psi(n,z) = \beta(z)\phi(n,z) + \alpha(z)\phi(n,z^{-1})$, $\phi(n,z) = -\beta(z^{-1})\psi(n,z) + \alpha(z)\psi(n,z^{-1})$.
5.	Asymptotic Behavior:	$\psi(n,k) \underset{\sim}{} \beta(k)e^{ink} + \alpha(k)e^{-ink}$ as $n \to +\infty$, $\zeta(n,z_j) \underset{\sim}{} c_j(z_j)^n$ as $n \to +\infty$.
6.	Analyticity Properties:	i) $\lvert\alpha(k)\rvert^2 = 1 + \lvert\beta(k)\rvert^2$, ii) $\alpha(z)$ is analytic for $\lvert z\rvert < 1$, continuous up to the boundary $\lvert z\rvert \leq 1$, except for poles at $z = \pm 1$. Its only zeros occur at the eigenvalues, $\alpha(z_j) = 0$ for $j = 1,\dots,N$.
7.	Scattering Data $\sum$:	$\sum = \{\beta(k)$ for $k \in [0,2\pi)$; z_j and c_j for $j = 1,\dots,N\}$.
8.	Evolution of $\sum$:	$\dot{\beta}(k) = 2i \sin k\, \beta(k)$, $\dot{z}_j = 0$, $\dot{c}_j = \frac{1}{2}(z_j - z_j^{-1})c_j$.

Table 2. Summary of scattering theory for the Toda lattice.

In the canonical theory, we coordinatize the scattering data $\sum$ by (θ, J) (see Table 2 for the notation):

(9)
$$\begin{aligned}
\theta(k) &= \arg \beta(k) \ , \\
J(k) &= \frac{1}{\pi} \sin k \ \ell n[1 + |\beta(k)|^2], \ 0 < k < \pi, \\
\theta_j &= \ell n \ \beta_j, \ \beta_j \equiv \left(-c_j^2 \ z_j \frac{d\alpha}{dz}\Big|_{z_j}\right), \\
J_j &= -(z_j + z_j^{-1}), \quad j = 1,\ldots,N.
\end{aligned}$$

Under this coordinatization of $\sum$, the inverse scattering transform can be interpreted as a canonical map from $\bar{S}$ with Poisson bracket characterized by C to $\sum$ with Poisson bracket characterized by J_Σ, that is

$$\{F,G\}_{J_\Sigma} \equiv \int_0^\pi dk \left[\frac{\delta F}{\delta\theta(k)} \frac{\delta G}{\delta J(k)} - \frac{\delta F}{\delta J(k)} \frac{\delta G}{\delta\theta(k)}\right] + \sum_{j=1}^N \left[\frac{\partial F}{\partial\theta_j} \frac{\partial G}{\partial J_j} - \frac{\partial F}{\partial J_j} \frac{\partial G}{\partial\theta_j}\right] .$$

In the rest of this section, we summarize the proof that this map is indeed canonical. Our procedure follows closely that of Zakharov and Manakov [34] for the nonlinear Schrödinger equation--that is, we show the Poisson brackets are preserved, $\{F,G\}_C = \{F,G\}_{J_\Sigma}$ for all scalar valued functions F and G.

Actually it is sufficient to verify the following:

$$\{J_j, J_{j'}\}_C = \{\theta_j, \theta_{j'}\}_C = \{J_j, J(k)\}_C = \{\theta_j, \theta(k)\}_C = \{J_j, \theta(k)\}_C = \{\theta_j, J(k)\}_C = 0 \ ,$$

(10)
$$\{\theta_j, J_{j'}\}_C = \delta_{jj'} \ ,$$

$$\{\theta(k), J(k')\}_C = \delta(k-k') \ .$$

We begin by setting $(L_0 f)(n) \equiv \frac{1}{2}(f_{n+1} + f_{n-1})$ and $L \equiv L_0 + V$. Consider an infinitesimal change in the potential from V to $V + \delta V$, which induces a change in the generalized eigenfunction from ψ to $\psi + \delta\psi$, and thus a change in its asymptotic behavior as described by the coefficients $\alpha(k)$ and $\beta(k)$:

$$(L-\lambda)\psi = 0 \Rightarrow (L-\lambda)\delta\psi = -(\delta V)\psi, \ \delta\psi(n,k) \to 0 \ \text{ as } \ n \to -\infty.$$

This inhomogeneous equation is solved by a <u>Green's</u> function, which gives the representation

$$\delta\psi(n,k) = \frac{i}{\sin k} \sum_{m=-\infty}^{n} \left[\psi(n,-k)\psi(m,k) - \psi(n,k)\psi(m,-k)\right] \left[(\delta V\psi)(m)\right] .$$

Comparing this expression as $n \to +\infty$ with the asymptotic behavior

$$\delta\psi(n,k) \sim \delta\beta(k)e^{ink} + \delta\alpha(k)e^{-ink} \quad \text{as} \quad n \to +\infty,$$

we obtain the key representations

$$\delta\alpha(k) = \frac{-i}{\sin k} I[\phi(k),\psi(k)],$$

(11)

$$\delta\beta(k) = \frac{i}{\sin k} I[\phi(-k),\psi(k)],$$

where ϕ is the second generalized eigenfunction defined in Table 2, and

$$I[f,g] \equiv \sum_m \left[f(m) \Big((\delta V g)(m)\Big)\right],$$

$$= \sum_m \left[f_m \delta a_{m-1}\, g_{m-1} + f_m\, \delta a_m\, g_{m+1} + f_m\, \delta b_m\, g_m\right] .$$

(Bilinear functionals of this type were introduced for p.d.e.'s in [1].) Using this formula for $I[\phi(k),\psi(k)]$, we compute from (11)

$$\frac{\delta\alpha(k)}{\delta a_m} = \frac{-i}{\sin k}\left[\phi_{m+1}(k)\ \psi_m(k) + \phi_m(k)\ \psi_{m+1}(k)\right], \tag{12}$$

and similar expressions for $\frac{\delta\alpha(k)}{\delta b_m}$, $\frac{\delta\beta(k)}{\delta a_m}$, and $\frac{\delta\beta(k)}{\delta b_m}$.

When using these to check the invariance of the Poisson brackets, one makes use of an identity involving Wronskians. Define the Wronskian by $W_n(f,g) \equiv a_n(f_{n+1}\, g_n - f_n\, g_{n+1})$. Then if $Lf = \lambda f$ and $L\tilde{g} = \tilde{\lambda}\tilde{g}$, one has

$$(\lambda - \tilde{\lambda}) f_n \tilde{g}_n = W_n(f,\tilde{g}) - W_{n-1}(f,\tilde{g}) , \tag{13}$$

which follows immediately from the difference equations.

Consider the Poisson bracket characterized by C between $\alpha(k)$ and $\beta(k)$,

$$\{\alpha(k),\beta(\tilde{k})\}_C = \begin{pmatrix} \nabla_a\alpha(k) \\ \nabla_b\alpha(k) \end{pmatrix} \cdot \begin{pmatrix} 0 & \frac{1}{4}C \\ -\frac{1}{4}C^T & 0 \end{pmatrix} \begin{pmatrix} \nabla_a\beta(\tilde{k}) \\ \nabla_b\beta(\tilde{k}) \end{pmatrix} .$$

Replacing $\nabla_a\alpha(k)$ with (12), making similar replacements for $\nabla_b\alpha(k)$, $\nabla_a\beta(\tilde{k})$, and $\nabla_b\beta(\tilde{k})$, and using the Wronskian identity (13) leads to a telescoping series for $\{\alpha(k),\beta(\tilde{k})\}_C$. Summing this series and using the asymptotic behavior summarized in Table 2 yields

$$\{\alpha(k),\beta(\tilde{k})\}_C = \frac{\sin^2 k + \sin^2 \tilde{k}}{8 \sin k \sin \tilde{k}(\cos \tilde{k} - \cos k)} \alpha(k)\beta(\tilde{k}) + \lim_{M\to+\infty} \frac{e^{2iM(k-\tilde{k})}}{\cos k - \cos \tilde{k}} \alpha(k)\beta(\tilde{k}) ,$$

where the limit as $M \to +\infty$ is interpreted in the sense of distributions. Specifically, the identity

$$\lim_{M\to+\infty} \frac{i}{2\pi} \frac{\sin k}{1 - \cos k} e^{-iMk} = \delta(k)$$

yields

$$\{\alpha(k),\beta(\tilde{k})\}_C = \frac{\sin^2 k + \sin^2 \tilde{k}}{8 \sin k \sin \tilde{k}(\cos \tilde{k} - \cos k)} \alpha(k)\beta(\tilde{k}) - \frac{i\pi}{\sin k} \alpha(k)\beta(k)\ \delta(k-\tilde{k}) .$$

Similarly, we obtain

$$\{\alpha(k),\alpha(\tilde{k})\}_C = 2\pi i\beta(0) \left(\frac{1 + \cos k}{\sin k} \alpha(k)\delta(\tilde{k}) - \frac{1 + \cos \tilde{k}}{\sin \tilde{k}} \alpha(\tilde{k})\delta(k)\right) ,$$

$$\{\beta(k),\beta(\tilde{k})\}_C = \frac{2\pi i}{\sin k} |\alpha(k)|^2\ \delta(k+\tilde{k}) .$$

Using these Poisson bracket identities together with the definitions of the (θ,J) variables (9) yields, for example,

$$\{\theta(k),J(\tilde{k})\}_C = \delta(k-\tilde{k}) - \delta(k+\tilde{k}).$$

Since $k \in (0,\pi)$, only the first delta function counts. (Notice that the constraint $J(-k) = -J(k)$, which follows from the definition $J(k) \equiv \frac{1}{\pi} \sin k\ \ell n[1 + \beta(k)\beta(-k)]$, allows us to consider k only in the range $(0,\pi)$.) The Poisson brackets for the remaining continuous variables follow similarly.

Turning to the discrete variables, we first assume $(a_n - \frac{1}{2})$ and b_n

to have compact support (in n). In this case $\beta(z)$ is an analytic function in the interior of the unit disc. Consider an eigenvalue z_j and an infinitesimal change from z_j to $z_j + \delta z_j$, which is induced by the infinitesimal change δV. To compute δz_j, recall $\alpha(z_j) = 0$; therefore,

$$0 = \alpha(z_j(V+\delta V)) - \alpha(z_j(V)) = \delta\alpha + \alpha'\delta z_j \Rightarrow \delta z_j = -[\alpha'(z_j)]^{-1}\delta\alpha(z_j).$$

Thus, $\text{grad}_{(a,b)} z_j$ is computed from formulas for $\text{grad}_{(a,b)}\alpha$ which we have already found; in particular,

$$\frac{\delta z_j}{\delta a_n} = \frac{-4\beta(z_j)}{\alpha'(z_j)[z_j - z_j^{-1}]}\,\phi_{n+1}(z_j)\phi_n(z_j) ,$$

$$\frac{\delta z_j}{\delta b_n} = \frac{-2\beta(z_j)}{\alpha'(z_j)[z_j - z_j^{-1}]}\,\phi_n^2(z_j) ,$$

where we have used the fact that at a bound state z_j, $\psi_n(z_j) = \beta(z_j)\phi_n(z_j)$. Using these formulas for grad z yields, for example,

$$\{\theta_j, J_{j'}\} = \delta_{jj'} .$$

The remaining Poisson brackets (10) follow similarly. Finally, to remove the restriction to compact support, one follows the computations outlined in reference [19].

3. Several Remarks Related to Canonical Structures

In this section we discuss several points, each of which is closely related to the canonical structure. Each point follows from a consideration of the function $\ell n\ \alpha(z)$ which satisfies the Poisson-Jensen identity

$$(14)\qquad \ell n\ \alpha(z) = -\ell n\ \alpha(0) + \frac{1}{\pi}\int_0^{\pi} dk\left[\ell n(|\alpha(k)|^2)\left(\frac{e^{ik}}{e^{ik}-z}\right)\right] + \sum_{j=1}^{N} \ell n\left[\frac{z-z_j}{z-z_j^{-1}}\right] ,$$

and has the power series expansion

$$(15)\qquad \ell n\ \alpha(z) = \sum_{p=0}^{\infty} C_p z^p , \quad |z| < \min\{|z_1|,\ldots,|z_N|\} .$$

Since $\alpha(z)$ is constant in t as (a,b) evolves according to the Toda equations, the family $\{C_p\}$ consists of constants of the motion (which can be expressed explicitly in terms of the scattering data $\sum$ by expanding the Poisson-Jensen formula).

On the other hand, the family $\{\tilde{C}_p\}$, which is defined by

$$\ell n\ \alpha(\lambda) = \sum_{p=0}^{\infty} \tilde{C}_p \lambda^{-p}, \quad \lambda \equiv \frac{1}{2}(z+z^{-1}) ,$$

also consists of constants of the motion; these are naturally given in terms of the (a,b) variables [9]:

$$\tilde{C}_p = \begin{cases} \ell n\ \alpha(\lambda=+\infty) = \ell n\ \alpha(z=0), & p = 0 , \\ \frac{1}{p}\ \mathrm{tr}(L^p - L_0^p) , & p \geq 1 , \end{cases}$$

where $\mathrm{tr}(A)$ represents the trace of the linear operator A. Of course, $\{C_p\}$ and $\{\tilde{C}_p\}$ are linear combinations of each other (since $\lambda = \frac{1}{2}(z+z^{-1})$).

We turn now to several remarks about the physical significance of these motion invariants.

Remark 1. $\ell n\ \alpha(z=0) = -\ell n\left[\prod_n 2a_n\right] = -\frac{1}{2}(Q_\infty - Q_{-\infty})$.

Thus, $\ell n\ \alpha(z=0)$ is proportional to the total elongation of the (Q,P) lattice from its equilibrium length. Moreover, these equalities show the total elongation $(Q_\infty - Q_{-\infty})$ to be a constant of the motion for the Toda equations.

This representation of $(Q_\infty - Q_{-\infty})$ can be verified in several ways. For example, from (11) we find

$$\delta\alpha(\lambda) = \pm \frac{1}{\sqrt{\lambda^2-1}}\ I[\phi(\lambda),\psi(\lambda)] ,$$

which, after a little calculation, yields (to leading term in λ)

$$\nabla_a \ell n\ \alpha(\lambda=+\infty) = -\frac{1}{a}, \quad \nabla_b \ell n\ \alpha(\lambda=+\infty) = 0 .$$

We integrate these differential equations with respect to the (a,b) variables to obtain

$$\ell n\ \alpha(\lambda{=}{+}\infty) = -\ell n \prod_n 2a_n + C\ ,$$

where the constant of integration C is independent of the variables of integration (a,b). To determine this constant, consider the free lattice ($a_n = \frac{1}{2}$, $b_n = 0$). In this case $\alpha(\lambda) = 1$, which implies $C = 0$. The rest of the representation is clear from the map taking (a,b) to (Q,P). (A second proof follows directly from the scattering theory as described by Case and Chiu [4].)

Remark 2. This representation of $(Q_\infty - Q_{-\infty})$ has some rather interesting physical consequences which may be deduced from the following facts (see below for proof):

i) If the reflection coefficient $R(k) = \beta(k)/\alpha(k) \equiv 0$, then $(Q_\infty - Q_{-\infty}) < 0$.

ii) If the point spectrum of L is empty, $(Q_\infty - Q_{-\infty}) > 0$.

iii) If $(Q_\infty - Q_{-\infty}) = 0$, L must have a nonempty point spectrum and R(k) cannot vanish identically.

(It is assumed that the lattice is not identically at rest.)

Physically, i) says that pure soliton states cause the lattice to contract. By ii), the "ringing components" of the wave, which arise from the continuous spectrum, cause the lattice to expand. By iii), all localized initial conditions, which automatically satisfy $(Q_\infty - Q_{-\infty}) = 0$ must evolve into a wave which contains both soliton and ringing components. Moreover, this happens no matter how weak the initial data.

To establish these three facts, we use the representation (14) for $\ell n\ \alpha(z{=}0) = -\frac{1}{2}(Q_\infty - Q_{-\infty})$ to write

$$(Q_\infty - Q_{-\infty}) = \frac{1}{\pi}\int_0^\pi dk\ \ell n\left[(1 - |R(k)|^2)^{-1}\right] + \sum_{j=1}^N \ell n\ z_j^2\ .$$

Now: i) If $R(k) \equiv 0$, then $(Q_\infty - Q_{-\infty}) = \sum_{j=1}^N \ell n\ z_j^2 < 0$ since $|z_j| < 1$.

ii) If L has no point spectrum, then $(Q_\infty - Q_{-\infty}) > 0$ since $(1 - |R(k)|^2)^{-1} > 1$.

iii) Follows similarly.

Remark 3. While $\tilde{C}_0$ describes the expansion or contraction of the lattice, $-2\tilde{C}_1$ gives its total momentum,

$$-2\tilde{C}_1 = -2\ \mathrm{tr}(L-L_0) = -2 \sum_n b_n = \sum_n P_n ,$$

which is a constant of the motion.

Remark 4. In order to construct additional completely integrable systems, one is tempted to use the constants $\tilde{C}_0$ and $\tilde{C}_1$ as Hamiltonians. However, since (by direct calculation) $\mathrm{grad}_{(a,b)}\tilde{C}_0$ and $\mathrm{grad}_{(a,b)}\tilde{C}_1$ span the null space of $\mathcal{C}$, these generate trivial flows on $\bar{\mathcal{S}}$.

For example, consider the total momentum $\tilde{C}_1$ which, on $\mathcal{S}$, generates translation of the center of mass. On the other hand, on $\bar{\mathcal{S}}$ it generates the trivial flow

$$\dot{\begin{pmatrix} a \\ b \end{pmatrix}} = \mathcal{C}\ \mathrm{grad}_{(a,b)}\tilde{C}_1 = \begin{pmatrix} 0 \\ 0 \end{pmatrix} .$$

In some sense $\mathcal{C}$ takes all center of mass motion to rest. The situation is worse than this, however. On $\sum$, the constant of motion $\tilde{C}_1$ is represented as

$$\tilde{C}_1 = \int_0^{\pi} dk[J(k)\cot k] + \sum_{j=1}^{N} (z_j - z_j^{-1}) ,$$

and since T_Σ is not singular, generates a nontrivial flow on $\sum$:

$$\dot{\theta}(k) = \frac{\delta C_1}{\delta J(k)} = \cot k, \quad \text{etc.}$$

Similar results apply for the constant of motion $\tilde{C}_0$. It generates a nontrivial flow on $\sum$, a trivial flow on $\bar{\mathcal{S}}$, and a very strange flow on $\mathcal{S}$ ($\dot{P}_n = 0$ for all n; $\dot{Q}_n = 0$ for all finite n; $\dot{Q}_{\pm\infty} = \mp \frac{1}{2}$).

This indicates the canonical nature of the map breaks down on certain parts of $\mathcal{S}$ or $\bar{\mathcal{S}}$; to remedy this, one must restrict the map's domain to submanifolds. In the case of $\bar{\mathcal{S}} \to \sum$, these are sets of the form $\tilde{C}_0$ = constant, $\tilde{C}_1$ = constant.

Even though $\tilde{C}_0$ and $\tilde{C}_1$ by themselves generate trivial flows on $\bar{\mathcal{S}}$, they will be useful when we identify an infinite family of completely integrable Hamiltonian systems.

Remark 5. Each member of the family $\{C_p, p \geq 2\}$ generates a Hamiltonian system which the inverse-scattering transform renders completely integrable. From the Poisson-Jensen identity (14), we can give a scattering data representation of C_p by [9]:

$$C_p = \frac{1}{\pi}\int_0^\pi dk\ \ln|\alpha(k)|^2 \cos pk + \sum_{j=1}^{N} \frac{1}{p}\left[z_j^p - z_j^{-p}\right], \quad p \geq 2 .$$

Notice that $\sin k$ is a factor of $(\cos pk - 1)$ if p is even and of $(\cos pk - \cos k)$ if p is odd. This fact, together with the definition $J(k) \equiv \frac{1}{\pi} \sin k \times \ln|\alpha(k)|^2$, indicates a clean choice for the Hamiltonians:

$$(16)\quad H_p = \begin{cases} C_p - 2C_0 = \displaystyle\int_0^\pi dk\ J(k)\left[\frac{\cos(pk)-1}{\sin k}\right] + \sum_{j=1}^{N}\left[\frac{1}{p}(z_j^p - z_j^{-p}) - \ln z_j^2\right], & p \text{ even}, \\ C_p - C_1 = \displaystyle\int_0^\pi dk\ J(k)\left[\frac{\cos(pk)-\cos k}{\sin k}\right] + \sum_{j=1}^{N}\left[\frac{1}{p}(z_j^p - z_j^{-p}) - (z_j - z_j^{-1})\right], & p \text{ odd}, \end{cases}$$

where $p \geq 2$.

Note that H_2 is the Toda Hamiltonian; while the next higher Hamiltonian H_3 turns out to be cubic in the momentum variables P_j. The higher systems depend upon still higher powers of P. Thus, this infinite family of completely integrable systems seems primarily of mathematical rather than physical interest, since we know of no physical lattices with dynamics of this type.

Remark 6. In this remark we summarize an alternative derivation of an infinite collection of completely integrable systems. This approach adapts the "Clarkson method" (as discussed in Section 3 of reference [12]) to the Toda lattice. Its advantages over the method of Zakharov, Faddeev, and Manakov used throughout the rest of this paper are (i) the key role of the "linearized dispersion relation" is brought to the foreground, (ii) it might handle systems somewhat more general than Hamiltonian systems--perhaps "SIT-like" systems, and (iii) it is based directly on the "squared eigenfunctions" [19] which seem to be rather basic to all aspects of the inverse method. (Readers only interested in the canonical framework may omit the rest of this remark as it is primarily intended for those familiar with the Clarkson method.)

The formulas for $\delta\alpha$ and $\delta\beta$ (11) give

$$\frac{dR}{dt} = \left[\frac{iI(\psi,\psi)}{\sin k\ \alpha(k)\beta(k)}\right] R ,$$

for the temporal evolution of the reflection coefficient $R(k;t)$. As it stands, this formula is valid for <u>arbitrary</u> evolutions of V; however, if a <u>condition for trivial integration</u>,

$$\left[\frac{iI(\psi,\psi)}{\sin k\ \alpha(k)\beta(k)}\right] = (2i \sin k)\Omega, \tag{17}$$

is satisfied, we can immediately integrate the above equation to obtain $R(k;t)$,

$$R(k;t) = R(k;0) \exp (2i \sin k)\Omega .$$

Here, of course, Ω is to depend upon k only and is independent of α, β, R, ψ, etc. Once $R(k;t)$ is known, inverse-scattering theory can be used to find $V(t)$ (since similar equations hold for the rest of the scattering data).

The condition for trivial integration (17) will be satisfied only by special flows $V(t)$. It turns out that the "linearized dispersion relation" for these special dynamical systems is given by $\omega(k) = 2 \sin k\ \Omega(k)$; thus, the linearized dispersion relation is seen to be the basis for this approach. We allow Ω to be any analytic function of the "eigenvalue" $\lambda = \cos k$.

To place the condition for trivial integration in a useful form, we first find an eigenvalue problem satisfied by the "squared eigenfunctions," $u_n \equiv \psi_n^2$, $v_n \equiv a_n\psi_n\psi_{n+1}$:

$$v_{n-1} + v_n + b_n u_n = \lambda u_n ,$$

$$\sum_{m=-\infty}^{n} \left[(a_m^2 - a_{m+1}^2)u_{m+1} + b_m(v_m - v_{m-1})\right] + a_{n+1}^2 u_{n+1} + a_n^2 u_n = \lambda v_n ,$$

or in shorthand form in terms of an operator $\mathcal{O}$,

$$\mathcal{O}\begin{pmatrix} u \\ v \end{pmatrix} = \lambda \begin{pmatrix} u \\ v \end{pmatrix}, \quad \begin{pmatrix} u_n \\ v_n \end{pmatrix} = \begin{pmatrix} \psi_n^2 \\ a_n\psi_n\psi_{n+1} \end{pmatrix} .$$

Using this eigenvalue problem, one can now rewrite the condition for trivial integration in the equivalent form

$$\begin{pmatrix} \dot{b} \\ 2\,\dot{\ell n}\,a \end{pmatrix} \cdot \begin{pmatrix} u \\ v \end{pmatrix} - \begin{pmatrix} A \\ B \end{pmatrix} \cdot \Omega(\mathcal{O}) \begin{pmatrix} u \\ v \end{pmatrix} = 0 ,$$

where

$$\begin{pmatrix} A_n \\ B_n \end{pmatrix} \equiv \begin{pmatrix} 2(a_n^2 - a_{n-1}^2) \\ 2(b_{n+1} - b_n) \end{pmatrix} .$$

Introducing the adjoint $\mathcal{O}^+$ of the operator $\mathcal{O}$, we obtain

$$\left[\begin{pmatrix} \dot{b} \\ 2\,\dot{\ell n}\,a \end{pmatrix} - \Omega(\mathcal{O}^+) \begin{pmatrix} A \\ B \end{pmatrix} \right] \cdot \begin{pmatrix} u \\ v \end{pmatrix} = 0 .$$

Thus, the condition for trivial integration will be satisfied provided the (a,b) evolve according to

$$\begin{pmatrix} \dot{b} \\ 2\,\dot{\ell n}\,a \end{pmatrix} = \Omega(\mathcal{O}^+) \begin{pmatrix} A \\ B \end{pmatrix} .$$

This is the key point in this remark. By their very construction, <u>all dynamical systems of this form are rendered completely integrable by</u> the inverse-scattering transform.

For each $n \geq 0$, the special choice $\Omega = \Omega_n \lambda^n$ generates a completely integrable system. For example, $\Omega = 1$ yields essentially the Toda lattice equations. It can be shown that each one of these systems is a Hamiltonian system of the form

$$\begin{pmatrix} \dot{a} \\ \dot{b} \end{pmatrix} = C \, \text{grad}_{(a,b)} \bar{H}_n , \quad n = 0,1,\ldots ,$$

where each Hamiltonian $\bar{H}_n$ is a linear combination of $\{H_2, \ldots, H_{n-2}\}$. Thus, this alternative approach <u>with analytic dispersion relations Ω</u> is equivalent to the approach used in the rest of the text.

4. Normal Modes for the Toda Lattice

The (Θ, J) coordinatization of $\sum$ is equivalent, under one additional canonical transformation, to the coordinatization

$$A_+(k) = \frac{1}{\sqrt{2\pi}} e^{2ik}\beta(k) ,$$

$$A_-(k) = \frac{-i}{\sqrt{2\pi}} 2 \sin k \, e^{-2ik} \frac{\ell n(1 + |\beta(k)|^2)}{\beta(k)} ,$$

$$A_{+j} = \Theta_j ,$$

$$A_{-j} = J_j .$$

In these coordinates, the Toda Hamiltonian takes the form

$$H = \int_0^{\pi} dk \; 2i \sin k \; A_+(k)A_-(k) + \sum_{j=1}^{N} \left[\frac{1}{2}(z_j^{-2}-z_j^2) + \ell n \; z_j^2\right] ,$$

and the equations of motion become

$$\begin{pmatrix} \dot{A}_+ \\ A_- \end{pmatrix} = T_\Sigma \; \text{grad}_{(A_+,A_-)} H .$$

There are striking analogies between these (A_+,A_-)-coordinates and the canonical description of the normal modes of the harmonic lattice (Table 1). In particular, only the discrete components really distinguish the spectral description of the Toda Hamiltonian from the Fourier description of the harmonic Hamiltonian. Of course, the scattering transformation, that is the map from the physical (Q,P) description of the Toda lattice to $\sum$, is nonlinear while the Fourier transform from the (Q,P) description of the harmonic lattice to its normal mode representation is linear.

If the vibrations of the Toda lattice are infinitesimally close to equilibrium, the dynamics of the Toda lattice reduces to that of the harmonic lattice. For such small amplitude oscillations, we let $(Q,P) = \varepsilon(\tilde{Q},\tilde{P})$, then

$$\begin{pmatrix} \dot{\tilde{Q}} \\ \dot{\tilde{P}} \end{pmatrix} = T \; \text{grad}_{(\tilde{Q},\tilde{P})} \tilde{H} + O(\varepsilon) ,$$

$$H = \varepsilon^2 \tilde{H} + O(\varepsilon^3) = \frac{\varepsilon^2}{2} \sum_n [\tilde{P}_n^2 + (\tilde{Q}_n - \tilde{Q}_{n-1})^2] + O(\varepsilon^3) .$$

To clarify the significance of the strange (A_+,A_-)-coordinates, we show that they reduce to the usual normal modes for the harmonic lattice as $\varepsilon \to 0$. This reduction will follow from the (discrete) integral equation for ψ,

$$\psi(n,k) = e^{-ink} - \frac{1}{\sin k} \sum_{m=-\infty}^{n} \sin[(n-m)k] \Big((V\psi)(m)\Big) ,$$

together with the integral representations of the scattering coefficients α and β:

$$\beta(k) = \frac{i}{2 \sin k} \sum_m e^{-imk} \Big((V\psi)(m)\Big) ,$$

(18)

$$\alpha(k) = 1 - \frac{i}{2 \sin k} \sum_m e^{imk} \Big((V\psi)(m)\Big) .$$

The representation for β yields

$$\beta(k) = \frac{i\varepsilon}{2 \sin k} \sum_m e^{-2imk} \frac{e^{ik}}{2}(\tilde{Q}_{m-2}-\tilde{Q}_{m-1}) + \frac{e^{-ik}}{2}(\tilde{Q}_{m-1}-\tilde{Q}_m) - \tilde{P}_{m-1} + O(\varepsilon^2) ,$$

which, together with the definitions in Table 1, yields

$$\beta(k) = \varepsilon\sqrt{2\pi}\, e^{-2ik}\, \tilde{\mathcal{Q}}(2k) - \frac{i}{\omega(2k)} \tilde{\mathcal{P}}(-2k) + O(\varepsilon^2) .$$

From this approximation, we obtain the approximations for the Toda variables[10]

$$A_+(k) \simeq \varepsilon\tilde{A}_+(2k) + O(\varepsilon^2); \quad A_-(k) \simeq 2\varepsilon\tilde{A}_-(2k) + O(\varepsilon^2) .$$

As the vibrations become infinitesimal, these canonical coordinates (A_+,A_-) tend to $(\tilde{A}_+,\tilde{A}_-)$, i.e. to a canonical description of the normal modes for the harmonic lattice. Moreover, in this limit the continuous variables $(A_+(k),A_-(k))$ provide a complete description of the linear lattice since the Fourier components are complete. (We have been considering the boundary conditions $(Q_n,P_n) \to (0,0)$ as $n \to \pm\infty$.) Thus, we know that the discrete components must contribute only at higher order in ε, and it is merely necessary to verify this fact.

For example, we show that the discrete components enter the Hamiltonian at a lower order in ε than the continuous components. Since $(Q_\infty-Q_{-\infty}) = 0$, the

---

[10] The factor of 2 compensates, $\int_0^\pi dk\, \omega(2k)A_+(k)A_-(k) \simeq \varepsilon^2 \int_0^{2\pi} dk'\, \omega(k')\tilde{A}_+(k')\tilde{A}_-(k') + O(\varepsilon^3)$.

solution must contain solitons as well as ringing components. In particular, at least one eigenvalue $\bar{z}$ must exist no matter how weak the (nonzero) initial data.

Realizing that this eigenvalue is a zero of α, $\alpha(\bar{z}) = 0$, we continue the summation representation (18) of α into the interior of the disc and find

$$\bar{z} = \underline{+}1 + E_1\varepsilon + E_2\varepsilon^2 + O(\varepsilon^3) .$$

We will not need any explicit representation of the coefficients $\{E_i\}$, although we do remark that E_1 is proportional to the total momentum of the lattice and, in general, will not be zero. From this form of $\bar{z}$, we find

$$\frac{1}{2}(\bar{z}^{-2} - \bar{z}^2) + \ell n\ \bar{z}^2 \underline{\sim} O(\varepsilon^3);$$

thus, the discrete components enter the Hamiltonian at a lower order in ε than the continuous components.

We emphasize two features of this linearization process. First, the (A_+,A_-)-coordinates provide a Hamiltonian description of the dynamics of the Toda lattice that is (i) linear, (ii) decoupled, (iii) explicitly dependent on the dispersion relation, and (iv) reduces to a Hamiltonian representation of the normal modes for the harmonic lattice. These features seem to justify the interpretation of the inverse-scattering method as a canonical transformation which maps the Toda lattice into normal mode form. The inverse-scattering transformation has identified the normal modes of oscillation for the (nonlinear) Toda lattice.

Secondly, this kind of linearization is not a correct solution technique for the Toda equations—even when the displacements are infinitesimally close to equilibrium. In effect, the linearized equations ignore the discrete components. Solitons are always present for localized disturbances; in fact, they are the most important waves in the Toda lattice. To ignore their presence, even when they are "weak and wide," introduces secular errors of order $(\varepsilon e^{\varepsilon t})$ into the approximation scheme.

5. The Constants of Motion

In this section we discuss in some detail two infinite collections of constants of the motion for the Toda lattice--the "action variables" $\{J\} = \{J(k), k \in (0,\pi) \mid J_j, j = 1,\ldots,N\}$ and the "higher constants" $\{C_p\}_{p=0}^{\infty}$.[11] The main point in this section is to provide a proof that these two families are equivalent. In addition, we comment on the physical information carried by these motion invariants and mention several mathematical questions common to such infinite families of constants of motion.

Throughout this paper we have frequently referred to families of constants of motion for the Toda lattice. There is clearly no such thing as "the" family; from any list of invariants, one can get a superficially different list by taking linear combinations or by performing more complicated (invertible) transformations. The families $\{C_p\}$ and $\{\tilde{C}_p\}$, defined, respectively, through $\ln \alpha = \sum C_p z^p$ and $\ln \alpha = \sum \tilde{C}_p \lambda^{-p}$, are so related. However, one can achieve a unique determination of the invariants by imposing enough structural requirements. For example, in reference [20], it is shown that there are no additional constants for the KdV equation, besides those shown to exist in that paper, which have local, polynomial, irreducible, isotropic, etc. densities.

Now it is rather difficult, even in principle, to see in what sense an infinite family of invariants, characterized by such structural properties, is "complete." Matters are simpler if the phase space has finite dimension (say $2n$). If a Hamiltonian system possesses n independent integrals in involution[12], then these constants may be introduced as action-type variables, with the n angular variables being obtained by quadrature [2]. Furthermore, in finite $(2n)$ dimensions, there are at most n independent integrals in involution, so one knows to look for <u>exactly</u> n of them; if n are found, integrability of the system is (in

---

[11] Recall: C_0, C_1, C_2 represent the elongation, the total momentum, and the total energy of the Toda lattice. The others, $\{C_p, p \geq 3\}$, are analogous to the higher polynomial constants for the KdV equation.

[12] Two constants of the motion, C_1 and C_2, are said to be in <u>involution</u> if their Poisson bracket vanishes, $\{C_1, C_2\} = 0$. They are <u>independent</u> if they have linearly independent gradients.

principle!) elementary.[13]

There is no general way, at present, to extend this argument to an infinite-dimensional phase space, that is, to decide when an infinite family of constants represents "half" the required variables. In the special case of the Toda lattice, the family $\{J\}$ consists of independent constants which are pairwise in involution; furthermore, the (Θ,J) coordinate system is of "action-angle" type--completeness being guaranteed by the inverse-scattering theory. Thus, the set $\{J\}$ is an example of a <u>maximal set of commuting integrals</u> (in ∞-dimensions).

The constants $\{C_p\}$ are also independent and pairwise in involution.[14] We will show that the family $\{C_p\}$ is "complete" by proving that $\{C_p\}$ is equivalent to $\{J\}$. Essentially, we indicate how the constants $\{C_p\}$ can be used to find all of $\{J\}$.[15] Such a verification of "completeness," which makes use of the very integrability it is supposed to imply, is quite different in spirit from the usual finite-dimensional arguments.

Once this equivalence has been established, the families $\{J\}$ and $\{C_p\}$ are known to carry the same physical information. Clearly the discrete action variables J_j fix the number and speeds of the soliton components in the wave. Now consider the continuous action variables $J(k)$. These reduce, for weak $O(\varepsilon)$ oscillations, to the action variables $\{\tilde{J}\}$ for the harmonic lattice (Table 1). Of course, this family of constants $\{\tilde{J}\}$ is the most familiar example of a maximal set of commuting variables (in ∞-dimensions). While the actual motion of the harmonic lattice consists of a continuous sum of these normal modes, it is conventional to regard each component in isolation. In this sense, $\tilde{J}(k)$ physically

---

[13]This argument would apply, for instance, to the periodic n-mass Toda lattice. The final quadratures are not at all simple, however.

[14]$\{C_p,C_q\}_C = \{C_p,C_q\}_{T_\Sigma} = 0$. The first equality follows from the invariance of the Poisson brackets; the second from the representation of the constants $\{C_p\}$ in terms of (Θ,J), see (16).

[15]That the family $\{J\}$ determines all of $\{C_p\}$ follows immediately from equation (16).

represents the amplitude of the kth mode of oscillation, $\mathcal{X}(k) = \sqrt{\tilde{J}(k)/\omega(k)} \cdot \exp[i\tilde{\Theta}(k)]$. In the case of the Toda lattice, an analogous interpretation is valid. Here the actual wave contains a continuous, though nonlinear, "superposition" of the "continuous" or "ringing" components--with the constant $J(k)$ thought of as the amplitude of oscillation of the kth mode. That $J(k)$ measures the amplitude of ringing seems reasonable since these continuous components decay algebraically in time t and behave almost as normal modes for the harmonic lattice in the asymptotic limit of large t. In particular, as t increases, we expect $\sqrt{J(k)/\omega(k)}$ to measure directly the amplitude of oscillation of the kth ringing mode. K. Miura [26] presents some rather striking numerical evidence which supports this idea that the continuous components are almost linear waves.

Since it is equivalent to the family $\{J\}$, the family $\{C_p\}$ also fixes the number and speeds of the solitons in the wave, together with the amplitudes of its ringing components, but (of course) provides no information as to the location of the solitons or the "phase of the ringing" (which depend on the angle variables $\{\Theta\}$). Thus, the physical information contained in the family $\{C_p\}$ has been identified very precisely, and it is consistent with the approximate information deduced from this family in many applications [3], [5], [21].

We turn to the question: "When does the family $\{C_p\}$ determine $\{J\}$?" Mathematically, this question splits into three--one of <u>existence</u>, one of <u>uniqueness</u>, and one of <u>construction</u>. The existence problem is to find conditions on a numerical sequence $\{a_p\}$ which will guarantee the existence of a solution $(\Theta(t),J)$ of the Toda lattice equations with the property that the family of constants $\{C_p\}$, for this solution, will equal the prescribed numerical sequence $\{a_p\}$. Such a numerical sequence will be called <u>admissible</u>. Not every sequence of numbers could be admissible. For example, $a_p = p^p$ is not; the series $\sum(p^p)z^p$ converges only for $z = 0$, whereas the series $\ell n\, \alpha = \sum C_p z^p$ must have a nonzero radius of convergence (and then very special singularities on that circle, etc.). It would be interesting to find necessary and sufficient admissibility requirements. This existence question can be formulated as a moment problem with certain constraints on the unknown mass distribution. These existence problems do not arise in applications, where the numerical sequences are found by evaluating the constants $C_p = C_p[Q,P] = C_p[a,b]$ on localized initial data.[16] This procedure guarantees admissibility. Suppose, then, that we are given the $\{C_p\}$ as computed from some state (Q,P) of the lattice. We know that a corresponding set $\{J\}$ exists; the problem is to find it. The series $\sum C_p z^p$ is known to converge in the circle $|z| < |z_1|$, where it represents $\ln \alpha(z)$. z_1 is the smallest eigenvalue of L (which directly yields the speed of the slowest soliton). Initially its value is unknown. Now, the radius of convergence,[17]

$$|z_1| = \left[\overline{\lim}\, |C_p|^{1/p}\right]^{-1} ,$$

determines the magnitude of z_1. One knows that z_1 is real, and to determine the sign, one must check whether the series converges or is infinite at $\pm |z_1|$. In this connection, it is important to observe that these are the only possibilities since the Poisson-Jensen identity fixes the type of singularities, $\ln(z-z_1)$. (Were the singularity like $(z-z_1)^{-1}$, the series would diverge at $(-z_1)$ as well, due to oscillations.) Having found z_1, we now consider the series,

$$\ln \alpha(z) - \ln(z-z_1) = \sum_{p=0}^{\infty} \left[C_p - \frac{1}{p!} \frac{\partial^p}{\partial z^p} \ln \left(\frac{z-z_1}{z-z_1^{-1}} \right) \Bigg|_{z=z_1} \right] z^p ,$$

which converges up to $|z_2|$, the next eigenvalue, etc. All eigenvalues are removed in a finite number of steps, and thereupon $\ln(1 + |\beta(k)|^2)$ is determined as the boundary value on $|z| = 1$ of the (known) function

$$\sum_p C_p z^p - \sum_{j=1}^{N} \ln \left(\frac{z-z_j}{z-z_j^{-1}} \right) .$$

[16] Scattering theory is not necessary to do this; the C_p are linear combinations of the $\tilde{C}_p$, and the latter, being traces of $(L - L_0)^p$, are polynomials in (a,b).

[17] If this radius of convergence equals 1, there are no solitons and $J(k)$ is determined by the series at $|z| = 1$.

In this manner the family $\{C_p\}$ determines $\{J\}$. Moreover, the analyticity properties of $\ell n\ \alpha(z)$ guarantee this determination is unique. Thus, $\{C_p\}$ is equivalent to $\{J\}$.

There is, of course, no way to implement these steps in practice. In fact it is seldom even possible to compute the entire sequence from a given state (Q,P). (The entire sequence can be computed from a pure N-soliton state [15].) Our observation is purely theoretical--although closely related to the approximations carried out in practice [3], [5], [21].

Some problems of mathematical interest remain open. In addition to the moment formulation of the existence question, there is the problem of developing a criterion for completeness of $\{C_p\}$, without reliance on the scattering solution. This would become a particularly interesting point, were one to find an equation with an infinite, yet incomplete, family of invariants.[18] Finally, analogous results must hold for KdV; however, our arguments do not carry over, as they would require analyticity of $\ell n\ \alpha(k)$ in a neighborhood of $k = \infty$.

ACKNOWLEDGMENT

We wish to thank Professors R. Hirota and H. Rund for a number of helpful conversations.

APPENDIX. CANONICAL NOTATION

This appendix contains some basic facts about Hamiltonian systems. Whereas it is customary to describe these via a differential 2-form (symplectic form), we find it more convenient to adopt the dual approach, and to work primarily with Poisson brackets.

Let M be a manifold with coordinates $(z_1,\ldots,z_{2n}) \equiv z$, and let $C(z)$ be a $2n \times 2n$ matrix function on M (strictly: a rank 2, covariant tensor field). The Poisson bracket of two scalar-valued functions, $F(z)$ and $G(z)$, is defined by

---

[18] There are many important finite-dimensional systems which can be proved to have too few integrals for an action-angle solution to be possible.

$$\{F,G\}_C \equiv \sum_{i,j} C_{ij} \frac{\partial F}{\partial z_i} \frac{\partial G}{\partial z_j} \equiv (\text{grad } F,\ C \text{ grad } G)\ .$$

We require that (dropping the subscript C)

i) $\{F,G\} = -\{G,F\}$, (anti-symmetry),

ii) $\{F,\{G,H\}\} + \{H,\{F,G\}\} + \{G,\{H,F\}\} = 0$, (Jacobi identity).

The first condition forces C to be skew-symmetric, $C^T = -C$; the second is an integrability condition whose precise consequences we shall not need.

A system of differential equations is called a Hamiltonian system with Hamiltonian H if it can be placed in the form

(A.1) $$\dot{z}_i = \{z_i, H\}\ .$$

An equivalent expression is

(A.2) $$\dot{z} = C(z) \text{ grad } H(z)\ .$$

In the most familiar example, $C = T = \begin{pmatrix} 0 & I \\ -I & 0 \end{pmatrix}$, and (A.2) reads

(A.3) $$\dot{z}_i = \frac{\partial H}{\partial z_{i+n}}\ , \quad \dot{z}_{i+n} = -\frac{\partial H}{\partial z_i}\ , \quad 1 \le i \le n\ .$$

It is customary to set $z_i = q_i$, $z_{i+n} = p_i$, which yields the Poisson bracket in the usual form

(A.4) $$\{F,G\} = \sum_{i=1}^{n} \left[\frac{\partial F}{\partial q_i} \frac{\partial G}{\partial p_i} - \frac{\partial F}{\partial p_i} \frac{\partial G}{\partial q_i} \right]\ .$$

According to Darboux' theorem, the general equation (A.2) can be brought (locally) into this standard form (A.3) by a change in variables, provided C is nonsingular. Nevertheless, in the text we find it more convenient to retain certain naturally arising coordinates, while resigning ourselves to a Poisson bracket more complicated than (A.4).

Consider now two manifolds, with coordinates z and w, both equipped with Poisson brackets given, respectively, by $C_1(z)$ and $C_2(w)$. A map ϕ, $z \to w = \phi(z)$, is called canonical if Poisson brackets are preserved,

(A.5) $$\{F(w),G(w)\}_{C_2(w)} = \{F(\phi(z)),G(\phi(z)\}_{C_1(z)}\ ,$$

for all functions F and G. It is a consequence of this invariance that the Hamiltonian form of a system of equations, (A.1) or (A.2), is preserved. In order to check the canonical nature of a map, it is sufficient to verify (A.5) for a set of 2n functions with linearly independent gradients.

We should note that if C is nonsingular, the above development is equivalent to one that proceeds from a symplectic form. For the Toda lattice, the C which occurs naturally is singular; we could remedy this defect by constraining the coordinates, but that would cause more inconvenience than the singularity of C.

In the text, we adapt these notions to an infinite-dimensional setting. That part of the development will be quite formal.

REFERENCES

[1] M.J. ABLOWITZ, D.J. KAUP, A.C. NEWELL, AND H. SEGUR, The inverse scattering transform-Fourier analysis for nonlinear problems, Studies in Appl. Math. 53 (1974), 249-315.

[2] V.I. ARNOLD AND A.A. AVEZ, Ergodic Problems of Classical Mechanics, W.A. Benjamin, Inc., New York, New York, 1968.

[3] Y.A. BEREZIN AND V.I. KARPMAN, Nonlinear evolution of disturbances in plasmas and other dispersive media, Soviet Physics JETP 24 (1967), 1049-1056.

[4] K.M. CASE AND S.C. CHIU, The discrete version of the Marchenko equations in the inverse scattering problem, Rockefeller Univ. Report (1974).

[5] R. DECK, F. HOPF, AND G.L. LAMB, JR., Univ. of Arizona, preprint (1975).

[6] B.A. DUBROVIN AND S.P. NOVIKOV, Periodic and conditionally periodic analogs of multisoliton solutions of the Korteweg-deVries equation, Z. Eksper. Teoret. Fiz. 67 (1974), 2131-2144.

[7] L.D. FADDEEV, The inverse problem in the quantum theory of scattering, J. Mathematical Phys. 4 (1963), 72-104.

[8] H. FLASCHKA, On the Toda lattice. I. Existence of integrals, Phys. Rev. B 9 (1974), 1924-1925.

[9] __________, On the Toda lattice. II. Inverse scattering solution, Progr. Theoret. Phys. 51 (1974), 703-716.

[10] __________, Discrete and periodic illustrations of some aspects of the inverse method, Dynamical Systems, Theory and Applications, Battelle Seattle 1974 Rencontres, J. Moser, ed., Springer-Verlag, New York, N.Y., 1975, pp. 441-466.

[11] H. FLASCHKA AND D.W. MCLAUGHLIN, Canonically conjugate variables for the Korteweg-deVries equation and the Toda lattice with periodic boundary conditions, Progr. Theoret. Phys., to appear.

[12] H. FLASCHKA AND A.C. NEWELL, Integrable systems of nonlinear evolution equations, same volume as [10], pp. 355-440.

[13] C.S. GARDNER, Korteweg-deVries equation and generalizations. IV. The Korteweg-deVries equation as a Hamiltonian system, J. Mathematical Phys. 12 (1971), 1548-1551.

[14] C.S. GARDNER, J.M. GREENE, M.D. KRUSKAL, AND R.M. MIURA, Method for solving the Korteweg-deVries equation, Phys. Rev. Lett. 19 (1967), 1095-1097.

[15] C.S. GARDNER, J.M. GREENE, M.D. KRUSKAL, AND R.M. MIURA, Korteweg-deVries equation and generalizations. VI. Methods for exact solution, Comm. Pure Appl. Math. 27 (1974), 97-113.

[16] R. HIROTA, private communication, 1974.

[17] H. HOCHSTADT, On the determination of Hill's equation from its spectrum, Arch. Rational Mech. Anal. 19 (1965), 353-362.

[18] A.R. ITS AND V.B. MATVEEV, On Hill's operator with finitely many gaps, Functional. Anal. Appl. 9 (1975), 65-66.

[19] D.J. KAUP, Closure of the squared Zakharov-Shabat eigenstates, Clarkson College, preprint (1975).

[20] M.D. KRUSKAL, R.M. MIURA, C.S. GARDNER, AND N.J. ZABUSKY, Korteweg-deVries equation and generalizations. V. Uniqueness and nonexistence of polynomial conservation laws, J. Mathematical Phys. 11 (1970), 952-960.

[21] G.L. LAMB, JR., Higher conservation laws in ultrashort optical pulse propagation, Phys. Lett. 32A (1970), 251-252.

[22] P.D. LAX, Integrals of nonlinear equations of evolution and solitary waves, Comm. Pure. Appl. Math. 21 (1968), 467-490.

[23] S.V. MANAKOV, On complete integrability and stochastization in discrete dynamical systems, Z. Eksper. Teoret. Fiz. 67 (1974), 543-555.

[24] H.P. MCKEAN, JR. AND P. VAN MOERBECKE, The spectrum of Hill's equation, Invent. Math., 30 (1975), 217-274.

[25] D.W. MCLAUGHLIN, Four examples of the inverse method as a canonical transformation, J. Mathematical Phys. 16 (1975), 96-99. Erratum to appear in J. Mathematical Phys. 16.

[26] K. MIURA, The energy transport properties of one-dimensional anharmonic lattices, Ph.D. Thesis, Dept. of Computer Science, Univ. of Illinois at Urbana-Champaign (1973).

[27] M. TODA, Waves in nonlinear lattice, Prog. Theoret. Phys. Suppl. 45 (1970), 174-200.

[28] _______, Studies on a nonlinear lattice, Phys. Rep. 18C (1975), 1-124.

[29] M. WADATI, The exact solution of the modified Korteweg-deVries equation, J. Phys. Soc. Japan 32 (1972), 1681.

[30] M. WADATI, H. SANUKI, AND K. KONNO, Relationships among inverse method, Bäcklund transformation and an infinite number of conservation laws, Progr. Theoret. Phys. 53 (1975), 419-436.

[31] H.D. WAHLQUIST, Bäcklund transformation of potentials of the Korteweg-deVries equation and the interaction of solitons with cnoidal waves, this volume.

[32] H.D. WAHLQUIST AND F.B. ESTABROOK, Bäcklund transformation for solutions of the Korteweg-deVries equation, Phys. Rev. Lett. 31 (1973), 1386-1390.

[33] V.E. ZAKHAROV AND L.D. FADDEEV, Korteweg-deVries equation: A completely integrable Hamiltonian system, Funkcional. Anal. i Prilozen 5 (1971), 18-27 (translation in Functional Anal. Appl. 5 (1972), 280-287).

[34] V.E. ZAKHAROV AND S.V. MANAKOV, On the complete integrability of the nonlinear Schrödinger equation, Teoret. Mat. Fiz. 19 (1974), 332-343.

[35] V.A. ZHELUDEV, On perturbations of the spectrum of the one-dimensional selfadjoint Schrödinger operator with periodic potential, Problems in Mathematical Physics, 4, M. Sh. Birman, ed., Leningrad Univ. (1970).

GPSR Compliance

The European Union's (EU) General Product Safety Regulation (GPSR) is a set of rules that requires consumer products to be safe and our obligations to ensure this.

If you have any concerns about our products, you can contact us on ProductSafety@springernature.com

In case Publisher is established outside the EU, the EU authorized representative is:

Springer Nature Customer Service Center GmbH
Europaplatz 3
69115 Heidelberg, Germany

Batch number: 09151018

Printed by Printforce, the Netherlands